Electrophysiology Measurements for Studying Neural Interfaces

Electrophysiology Measurements for Studying Neural Interfaces

Mohammad M. Aria, PhD candidate in Biomedical engineering

Research Assistant, Koc University, Istanbul, Turkey

ACADEMIC PRESS
An imprint of Elsevier

Academic Press is an imprint of Elsevier
125 London Wall, London EC2Y 5AS, United Kingdom
525 B Street, Suite 1650, San Diego, CA 92101, United States
50 Hampshire Street, 5th Floor, Cambridge, MA 02139, United States
The Boulevard, Langford Lane, Kidlington, Oxford OX5 1GB, United Kingdom

Notices
Knowledge and best practice in this field are constantly changing. As new research and
experience broaden our understanding, changes in research methods, professional
practices, or medical treatment may become necessary.

Practitioners and researchers must always rely on their own experience and knowledge in
evaluating and using any information, methods, compounds, or experiments described
herein. In using such information or methods they should be mindful of their own safety
and the safety of others, including parties for whom they have a professional
responsibility.

To the fullest extent of the law, neither the Publisher nor the authors, contributors, or
editors, assume any liability for any injury and/or damage to persons or property as a
matter of products liability, negligence or otherwise, or from any use or operation of any
methods, products, instructions, or ideas contained in the material herein.

Library of Congress Cataloging-in-Publication Data
A catalog record for this book is available from the Library of Congress

British Library Cataloguing-in-Publication Data
A catalogue record for this book is available from the British Library

ISBN: 978-0-12-817070-0

For information on all Academic Press publications visit our website at
https://www.elsevier.com/books-and-journals

Publisher: Joe Hayton
Acquisitions Editor: Natalie Farra
Editorial Project Manager: Mona Zahir
Production Project Manager: Sreejith Viswanathan
Cover designer: Alan Studholme

Typeset by TNQ Technologies

This book is dedicated to my family:
My mother, for all your supports and care, I love you!
My father, for all your supports and encouragements,
you are always in my heart.
My beautiful sisters Sara and Samira who encouraged me
to finish this book!

Contents

Preface

The multidisciplinary field of bioelectronics and neuroprosthetics demand wide technical knowledge from a variety of scientific backgrounds, including electrical engineering, material sciences, biology, and neuroscience. Furthermore, recent advances in neural interfaces with bionanomaterials require precise investigation and understanding of the underlying neural stimulation mechanisms. For this important goal, all the elements necessary for such a project including the development of a neural interface, the design of experimentation with neuron-like cells or neurons, and the examination of the performance of neural interfaces with the aforementioned cells have been discussed in this book.

This book has six chapters including the fundamentals of bioelectricity and excitable membranes (Chapter 1), exploring neural stimulation techniques such as the photoelectrical and photothermal stimulation of neurons by introducing whole-cell patch-clamp electrophysiology (Chapter 2), electrophysiological studies of neuron-like cell lines (such as PC12, Neuro-2a, NG108, and SHSY) with data analysis of patch-clamp results from various relevant experiments and cell culture tips (Chapter 3), extracellular recording and spike sorting (Chapter 4), optical monitoring and the control of neuronal activities through fluorescent indicators and optogenetic technology (Chapter 5), and the operation of electronic circuits in the patch-clamp system (Chapter 6).

Whole-cell patch-clamp electrophysiology has been widely used to measure the electrophysiological properties (membrane potential and ionic conductances) of cell membranes that control neuronal functions such as firing of action potential or neural silencing. Therefore, it is a very useful technology for investigating the performance of neural interfaces. For example, photovoltaic interfaces that are embedded with organic polymers—such as poly(3-hexylthiophene)—organic pigment photocapacitors, and inorganic quantum dots for photoelectrical stimulation of cells have recently attracted attention for neuroprosthtics application, such as retinal prothesis. However, the study of the efficacy and the mechanism behind neural stimulation requires an analysis and understanding of cells' neurophysiological behaviors that have been realized by the patch-clamp technique. Hence, in Chapter 2, basic functions of the patch-clamp setup and its applications for studying various types of interfaces have been reviewed and working mechanisms of different types of photoelectrical and photothermal neural stimulation have been discussed. Although whole-cell patch-clamp and extracellular recording mainly used in studying neural stimulation methods, optical techniques provide remote neural modulation and monitoring based on optogenetics and fluorescent imaging. In this regard, I have reviewed the basics of calcium imaging and optical electrophysiology with some practical examples from recent studies in Chapter 5.

I tried my best to explain my experiences and recent studies in bioelectronic interfaces for neural stimulation in this book. And I hope that this book will be practical for new researchers to use its contents in their projects and studies. Enjoy it!

Mohammad Mohammadi Aria
April 2020

Bioelectricity and excitable membranes

1.1 Introduction

Electrical signals are generated, propagated, and processed in the brain. To implement different functions, the brain uses different neuronal circuits including different types of neurons and connections. The author begins with retinal cells as some of the most interesting neurons and their functions, together with photoreceptors that enable vision. The visual system includes retinal cells, photoreceptors, and visual cortex neurons. In fact, photoreceptors play the role of sensors that receive visual information from the environment. This information will be transmitted in a complex network of different types of neurons to the visual cortex with optic nerve. The visual cortex also has many different types of neurons, and they form different circuits to implement different functions. To begin with the basis of a neuronal circuit, one needs to understand elements of a neuronal circuits and how they work together to implement a function. More importantly, each neuron has voltage-gated ion channels or excitable membranes that enable them to receive and send electrical signals. The basic concept of excitable membranes and voltage-gated ion channels in cell physiology is the central focus of this chapter.

Decoding the content of neuronal signals is a major goal of neurobiological research. As different neuronal circuits have different roles, the meaning of the signals depends on their origin and where they are transmitted, as well as signal parameters such as the frequency and duration of activation. Each individual nerve cell can receive thousands of inputs from other neurons. In this way, with integrating this input information, the cell generates a new output message that can send a new complex meaning, such as the presence of light with different colors in one's field of vision.

This chapter explains how neuronal activities from a big picture (function of a group of cells in a circuit) to a small picture of a single neuron enable specific functions. In the following, basic components of a neuron including cell body, axons, and dendrites are discussed in detail. As neuronal signals and functions are transmitted and controlled by electrical signals, the concept of excitable membranes and a superfamily of voltage-gated ion channels has been explained to prepare audiences for the needs of neuronal stimulation and inhibition for neuronal prosthesis. The high importance of ion channels is because of their role in membrane excitability,

Electrophysiology Measurements for Studying Neural Interfaces. https://doi.org/10.1016/B978-0-12-817070-0.00001-4

encoding neuronal activity, their diversity, and their effects on cellular functions which need close attention for studying neuronal prosthesis. A second goal in this chapter is to understand the electrical circuit of a single neuron based on the Hodgkin-Huxley (H-H) quantitative description. The landmark work of the H-H model describes a quantitative and an electrical circuit to understand neuronal activities. This model correlates different elements of ionic conductance, cell membrane capacitance, and excitation amplitude to action potential shape in time.

There are two types of electrical signals in the brain: local graded potentials, which are localized over short distances, and action potentials, which are transmitted over long distances. For instances, in synapses (junctions of one neuron to another), upon release of chemical transmitter molecules that bind to specific chemoreceptor molecules in the target cell membrane, a local graded potential activates or inhibits the cell depending on the transmitter and the corresponding receptors. The efficacy of synaptic transmission may be affected by molecules, hormones, and drugs. Action potentials are also generated when the membrane potential is increased and passes a threshold level of -45 mV. In addition, action potentials could be like a train of spikes transmitted through myelinated axons.

1.2.1 Neuron cell

Each neuron cell consists of a cell body, an axon, and dendrites. Fig. 1.1 shows a ganglion neuronal cell structure. The cell body consists of the cell membrane, nucleus, and organelles like other type of cells. Axons are connections that go from one cell body to another targeted cell. Dendrites act as receivers for excitation and inhibition through incoming fibers. Not all neurons have similar structure, as shown in Fig. 1.1. Certain nerve cells without an axon can communicate through other neurons that have an axon. In this case, having dendrites enables neurons conducting impulses to target cells. While ganglion cells include dendrites, a cell body, and an axon, other cells such as photoreceptors do not have an axon or dendrites. Because the stimulation or in other words the input signal in photoreceptors comes through light, not through input from another neuron, they do not need to have axons or dendrites.

1.2.2 Signaling in nerve cells

The first biopotential to be discussed is resting membrane potential, which arises as a result of ionic charge balance across the cell membrane. Resting membrane potential is a negative value about -69 mV for neurons. It is measured from inside of the cell in comparison to the extracellular medium. The signals that increase the cell membrane potential to a more positive value lead to cell depolarization, making the inside less negative, while the signals that hyperpolarize it make the intracellular medium more negative. As mentioned, there are two different electrical signals. The first includes local graded potentials such as receptor potentials and subthreshold membrane potential oscillations. For example, light falling on a photoreceptor in the eye

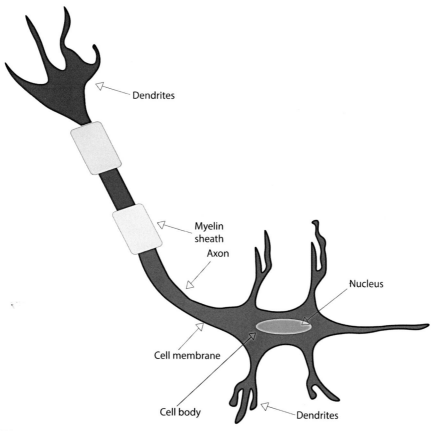

FIGURE 1.1

Structure of a typical neuron including cell body and membrane, nucleus, axon, myelin sheath, and dendrites.

leads to a hyperpolarization of the photoreceptor cell, which causes bipolar cell depolarization. This is because of the inhibitory mechanism of the synapse between them. This depolarization consequently excites the following cell, a ganglion cell, causing it to generate action potentials at a higher frequency. Local potentials vary in amplitude, depending on the strength of the activating signal. They usually spread only a short distance from their site of origin because they depend on passive electrical properties of the nerve cells. Action potentials are the second category of electrical signals that play an important role in signal communication and processing in neuronal networks if localized graded potentials are large enough to depolarize the cell membrane beyond a threshold level (called the threshold). Once action potentials are initiated, myelinated axons transmit them rapidly over long distances. For example, generated action potentials from ganglion cells are transmitted through axons to the optic nerve from the eye to the lateral geniculate nucleus, and then to the cortex. Action potentials have a constant amplitude and duration. The only properties that could be modulated are duration of activation, silencing, and refractory period of frequency.

1.2.3 Retinal structure

The different types of neuronal cells in the retinal layer include bipolar cells, ganglion cells, horizontal cells, retina amacrine cells, and rod and cone photoreceptors. Fig. 1.2 shows simply the structure of the vertebrate retina. Retinal structure includes photoreceptors as light sensors. A wide spectrum of light enters the eye and reaches the photoreceptors. These photoreceptors generate graded potentials, and bipolar cells transduce these graded potentials to the proper stimulation signals for triggering ganglion cells. In the dark condition, a photoreceptor (rod/cone) cell inhibits (hyperpolarizes) the ON bipolar cells, while it stimulates (depolarizes) the OFF bipolar cells by releasing glutamate. Under light conditions, the entering light produces graded potentials via the photoreceptor leading to hyperpolarization. On the other hand, retinal ganglion cells transmit visual information from the explained retina layer with propagation of action potentials through axons to several regions including the midbrain and diencephalon. These regions will process visual information at the end. One of the most important lessons of this part was that the position of neurons and their origin and the destination in the neural network provide valuable information about their functionality.

Damage to these cells as a result of aging and disease negatively affects vision and may even lead to blindness. One of the most important aspects of retinal prosthesis in recent decades has been the replacement of damaged photoreceptors with artificial photoreceptors. In recent years, researchers developed semiconductor-based interfaces for electrical transduction of light for retinal prosthesis application [1]. This semiconducting layer produces photoinduced potentials to trigger neurons. In Chapter 2, different types of neural interfaces embedded with semiconductor materials will be described. In brief, these interfaces upon light illumination can generate local photoinduced potentials or currents at their surfaces, which are the source of neural stimulation. Fig. 1.2B shows an animation that photoreceptors have been replaced by artificial semiconductor materials that absorb light in the visible spectrum. However, there are two different ways to implant the semiconductor interfaces: direct connection with the retina (epiretinal implant) or behind the retinal layer (subretinal).

1.3 Cell structure

1.3.1 Cell body

The spherical region near the center of the cell where the nucleus is located is called the cell body or soma. Similar to other cells, the cell body has the nucleus, nucleolus, endoplasmic reticulum (ER), Golgi apparatus, mitochondria, ribosomes, lysosomes,

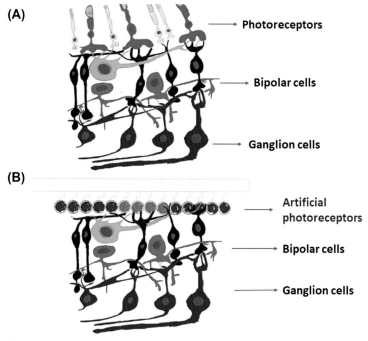

FIGURE 1.2

Schematic diagram of the retina with natural and artificial photoreceptors. (A) The whole retinal structure including photoreceptors, bipolar cells, and ganglions. (B) The whole retina structure including artificial photoreceptors, bipolar cells, and ganglions.

endosomes, and peroxisomes. These elements are generally responsible for different cell functions such as expression of genetic information, controlling the synthesis of cellular proteins (for instance, ER), involved in energy production, and cell growth. Fig. 1.3 shows the schematic of a cell with its elements and an excitable membrane. An excitable membrane coordinates ionic transport across the cell membrane based on the change in membrane potential. Excitability of the membrane is originated from the voltage-dependent electrical conductance of the ion channels including sodium, potassium, and calcium.

1.3.2 Excitable membranes

As Fig. 1.4 shows, the plasma membrane plays an important role in ionic transport as it is embedded with ion pores (channels) and pumps to transport ions. Voltage-gated ion channels, which are the most important element in electrical excitability of the cell membrane, have a great diversity and a large family which will be discussed in Section 1.4. These channels are controlled by the membrane potential. During the action potential generation, their opening and closing are coordinated by the

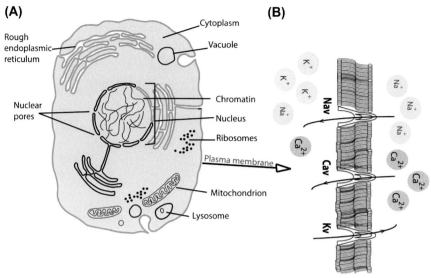

FIGURE 1.3

Diagram of a cell body structure and plasma membrane. (A) Cell body structure and
(B) plasma membrane structure including lipid bilayer and voltage-gated ion channels.

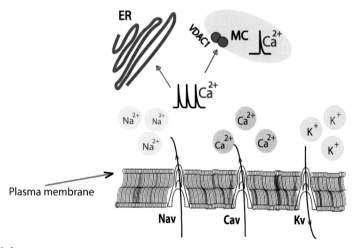

FIGURE 1.4

Schematic of plasma membrane and cell signaling through voltage-gated ion channels.

nonlinear behavior of cell conductance. More importantly, intracellular signaling
such as calcium level is modulated by ionic activities and membrane potential. Calcium plays two roles: as a charge transporter and an intracellular messenger. In fact,
during depolarization, calcium level inside the cell increases, and so during the burst

of action potentials, the changes in the intracellular calcium can be used to analyze neuronal activities as intracellular Ca^{2+} signaling can be changed during cell depolarization and action potential. Fig. 1.4 shows the intracellular calcium signaling between different elements of mitochondria and ER [1]. Ca^{2+} signals are generated through internal calcium stores and external sources of calcium (calcium ion channels and pumps). Voltage-gated ion channels are one of the important external sources (Section 1.4.2). It also affects ER-mitochondria interface, which is fundamental for different functional outcomes, such as cell metabolism or induction of cell death.

1.3.3 Equilibrium potentials and the Nernst equation

In the steady-state condition, cells have a biopotential, which is called resting membrane potential. Resting membrane potential is originated from different equilibrium or Nernst potential of voltage-gated ion channels. As the cell membrane separates intracellular ions from the extracellular medium, because of the difference in the ionic concentration, the Nernst potentials are created. Although there is different permeability to different ions in the cell membrane, the equilibrium condition may vary from one ion to another. At equilibrium condition, the effective influx and outflux through membrane is almost zero. In this condition, the diffusion force (as a result of thermal forces) induced by the gradient of ionic concentration is just opposite of the induced electrical force generated by electrical charge difference in both sides. Therefore, at equilibrium, we consider that electrical and diffusion forces are equal and lead to balance and equilibrium. As Fig. 1.5 shows, a cell membrane embedded by pores is permeable only to K^+ ions separating a high concentration of a salt K_A (A for anion) in the left side and a low concentration into the right side. A voltmeter is used to measure the potential across the cell membrane. At the

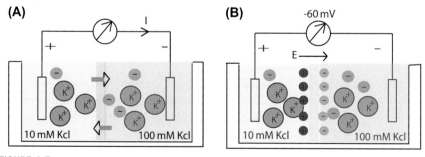

FIGURE 1.5

Block diagram of a model describing Nernst equation. (A) After recombination of high and low concentration of KCl because of the difference in the ionic concentration, there is a diffusion force and so an ionic current flow starts, and (B) after the charge transfer, constant ions across the membrane prevent any more diffusion current and so a built-in potential is created which could be calculated by the Nernst equation.

beginning that the diffusion starts from the high concentration to the low concentration side, the voltmeter shows 0 mV. However, during the diffusion of K^+ ions into the right-hand side, because of the accumulation of excess positive charges, an electrical potential will be created across the cell membrane. This electric field prevents anions from crossing the cell membrane, so it leads to halting charge separation. The electric field is built up before reaching an equilibrium value of E_K, where the electrical and diffusional forces are equal and there will not be any more transfer of potassium ions. To find an equation to calculate E_K, the first method is to use the Boltzmann equation of statistical mechanics. Based on the Boltzmann statistic, the relative probabilities at equilibrium of finding a charge in state 1 or in state 2 depends on the energy difference between these states, which is $u_2 - u_1$:

$$\frac{P_2}{P_1} = \exp\left(-\frac{u_2 - u_1}{kT}\right) \tag{1.1}$$

where k is the Boltzmann's constant and T is absolute temperature on the Kelvin scale. In this equation, the equilibrium distribution of charges can be found. This equation also shows that the probability of finding the charges in higher energy levels is lower than finding them in higher energy levels. In Eq. (1.1), c_2 and c_1 as the ionic concentrations are replaced with P_2 and P_1:

$$\frac{c_2}{c_1} = \exp\left(-\frac{u_2 - u_1}{RT}\right) \tag{1.2}$$

where R is the gas constant ($R = kN$). $U_1 - U_2$ is the molar energy difference and can be obtained as following:

$$U_1 - U_2 = RT \ln\frac{c_2}{c_1} \tag{1.3}$$

Considering mole of an arbitrary ion S has a charge of Z_s, then $U_1 - U_2$ becomes $Z_s F (E_1 - E_2)$. To obtain the equilibrium potential E_s as a function of the concentration ratio and the valence, from Eq. (1.3), we have the well-known Nernst equation:

$$E_S = E_1 - E_2 = \frac{RT}{Z_S F} ln \frac{[S]_2}{[S]_1} \tag{1.4}$$

From a thermodynamic perspective, at the equilibrium condition, the electrochemical potential of ion S on both sides is equivalent, and so the work to transfer a tiny charge of S from the left side to the right should be zero. This work includes two terms: the first term is $-RT \ln (c_2/c_1)$, which is the diffusion force, and the second term is $U_1 - U_2$, which is the electrical term. In the Nernst equation, ionic equilibrium potentials linearly depend on the absolute temperature and logarithmically on the ionic concentration ratio. As expected from Fig. 1.5, the sign of equilibrium potentials changes if the charge or direction of the gradient is reversed, and it becomes zero when there is no gradient. In case of a cell membrane, left side, which has low amount of potassium, is considered as inside (intracellular), the second side as outside (extracellular), and the whole cell membrane potential is measured inside minus outside. Other equilibrium potentials for other ions are obtained like K$^+$ ions:

$$E_k = \frac{RT}{F} \ln\left(\frac{P_K[K]_e}{P_K[K]_i}\right) \tag{1.5}$$

$$E_{Na} = \frac{RT}{F} \ln\left(\frac{P_{Na}[Na]_e}{P_{Na}[Na]_i}\right) \tag{1.6}$$

$$E_{Ca} = \frac{RT}{2F} \ln\left(\frac{P_{Ca}[Ca]_e}{P_{Ca}[Ca]_i}\right) \tag{1.7}$$

$$E_{Cl} = \frac{RT}{F} \ln\left(\frac{P_{Cl}[Cl]_e}{P_{Cl}[Cl]_i}\right) \tag{1.8}$$

where subscripts e and i indicate extracellular and intracellular, respectively. All these numbers are separately showing the equilibrium condition for membranes permeable to specific ions. This shows that if the membrane potential is kept at the equilibrium potential, there will be a net zero ionic current. One could ask, what if a membrane is permeable to different types of ions? How to calculate the resting membrane potential? The answer to this question is the Goldman-Hodgkin-Katz voltage equation:

$$E_m = \frac{RT}{F} \ln\left(\frac{P_{Na}[Na]_{out} + P_K[K]_{out} + P_{Ca}[Ca]_{out} + P_{Cl}[Cl]_{out}}{P_{Na}[Na]_{in} + P_K[K]_{in} + P_{Ca}[Ca]_{in} + P_{Cl}[Cl]_{in}}\right) \tag{1.9}$$

where E_m is the membrane potential and P_i is the permeability to a specific ion of i. In fact, the permeability of an excitable membrane to different ions varies as it is embedded with voltage-gated ion channels. However, in this equation, it includes the permeability of ion channels near the resting potential. Table 1.1 represents ionic concentrations, equilibrium potentials, and their relative permeability at 20°C.

1.3.4 Current-voltage relations of channels

To understand and study the electrophysiological behavior of excitable membranes, quantitative description of the ion channels is useful. A simple electrical passive RC model (with a battery resembling built-in resting potential) is widely used for the

Table 1.1 Ionic concentrations and the permeabilities of different ion channels.

V	[Ion]$_{inside}$ (mM)	[Ion]$_{outside}$ (mM)	E_{ion} (mV)	P_{ion}
Potassium	140	5	−89	1
Sodium	15	150	+61.5	0.04
Chloride	10	120	−66	0.45

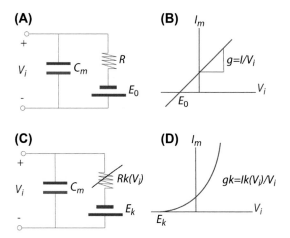

FIGURE 1.6

I-V model behavior of a passive cell. (A) RC passive model of cell membrane in which C_m is the cell membrane capacitance, R_m is the membrane conductance, and E_0 is the resting membrane potential. (B) I-V curve for the RC equivalent circuit model. (C) Electrical equivalent circuit model for a potassium ion channel. (D) I-V characteristic of a potassium ion channel showing a nonlinear and rectifying behavior.

passive cell with membrane potential less than −45 mv (the threshold for firing of action potential). In this model (as shown in Fig. 1.6), the membrane is considered having a capacitor because of the lipid bilayer and the conductance as a result of ion channels. The battery has resting potential of E_m and the net driving force on ion channels is $E_i − E_m$. As an approximation, this linear Ohm's law works fine (Fig. 1.6B), and after exciting of ion channels, when many pores are open, a nonlinear current-voltage relation will appear (Fig. 1.6D).

The linear I-V behavior can be easily calculated based on the Ohm's law:

$$I_m = g(Vi − E_0) \tag{1.10}$$

where I_m is the membrane current, g is the conductance ($g = 1/R$), and V_i is the input applied potential. On the other hand, the I-V relationship for a cell membrane with potassium voltage-gated channel is nonlinear as following:

$$I_K = G_K(V) \times (Vi − E_K) \tag{1.11}$$

where G_K is a nonlinear voltage-dependent conductance as a result of activated ion channels, and the $(V_i − E_k)$ is the applied force on the ion channels.

1.4 Superfamily of voltage-gated ion channels

There are more than 350 different ion channels expressed in the mammalian brain and 143 of them are voltage-gated. Gene expression, intracellular calcium alternation, and signal processing in neurons are some examples that rely on the activity of voltage-gated ion channels embedded in the cell membrane. They play crucial roles in cellular functions, communications, and electrical excitability. Gene expression, intracellular calcium alternation, and signal processing in neurons are some examples of functions that rely on the activity of voltage-gated ion channels embedded in the cell membrane. The superfamily of voltage-gated ion channels regulates all these physiological processes. This superfamily is made up of eight families of voltage-gated sodium, calcium, and potassium, calcium-activated potassium channels, potassium channels rectifying inwardly, cyclic nuclide-modulated ion channels, possible transient receptor channels, and two-pore potassium channels [2]. Fig. 1.7 illustrates the working mechanism of voltage-gated ion channels and the unrooted tree of the superfamily. The working mechanism of the ion channels could be simply described as when the ion channel is open or active, pores allow ions to pass through the cell membrane (as shown in Fig. 1.7A). When they close, the ionic transmission will be stopped. Controlling ionic fluxes in neurons facilitates encoding, process, and transmission of neuronal signals (action potentials) by these channels; and in the heart, Purkinje cells' ionic impulse propagation to numerous sites in the left and right ventricle facilitates contraction of the heart.

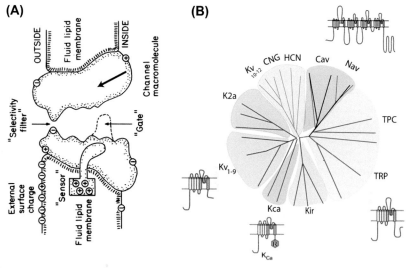

FIGURE 1.7

Voltage-gated ion channels and their working mechanism and concept. (A) The schematic of a voltage-gated ion channel, based on studies of voltage-gated sodium and voltage-gated potassium channel and (B) unrooted tree depicting VGIC superfamily members [2]. Subfamilies (clockwise) include voltage-gated calcium and sodium channels (Ca$_V$ and Na$_V$), two-pore (TPC) and transient receptor potential (TRP) channels, inwardly rectifying potassium channels (K$_{ir}$), calcium-activated potassium channels (K$_{Ca}$), voltage-gated potassium channels (K$_V$1–K$_V$9), K2P channels, voltage-gated potassium channels from the EAG family (K$_V$10–K$_V$12), cyclic nucleotide-gated channels (CNG), and hyperpolarization-activated channels (HCN) [3].

The structural architectures of ion channels consist of four variations built on a common structural theme that shapes the pore. Structural analysis of the voltage-gated ion channels has been done with high-resolution scanning electron microscope, and image reconstruction shows that there are four principal homologous domains (I to IV) that surround a central pore indicating entry ports that are laterally oriented in each domain for ion transit toward the central pore. All these domains are built from subunits that are thought to form membrane-spanning α-helices.

1.4.1 Voltage-gated sodium channels

Voltage-gated sodium (Na^+) channels are the main components contributing in the cell membrane excitability as the sodium influx leads to the cell membrane depolarization, so action potential generations and propagation in the axons. In this regard, they play an important role in the cell membrane excitability and electrical conductivity in neural networks. In addition, they have a critical role in controlling several pain syndromes, including inflammatory pain, neuropathic pain, and central pain associated with spinal cord injury. These channels close rapidly upon repolarization or more slowly on sustained depolarization. As shown in Fig. 1.8, the voltage-gated Na^+ channels are consisting of an α-subunit including of four homologous domains (DI−DIV) and an auxiliary β-subunit [4].

There are many different types of voltage-gated Na^+ channels. Na_v channel consists of the α-subunit and the channels with the different α-subunit subtypes, which are nine different α-subunits of $Na_v1.1$, $Na_v1.2$, $Na_v1.3$, $Na_v1.4$, $Na_v1.5$, $Na_v1.6$, $Na_v1.7$, $Na_v1.8$, and $Na_v1.9$. These channels are encoded by the genes SCN1A, SCN2A, SCN3A, SCN4A, SCN5A, SCN8A, SCN9A, SCN10A, and SCN11A, respectively. They are different in thresholds, inactivation and recovery from inactivation, different kinetic behaviors and voltage dependence, and sensitivity to tetrodotoxin (TTX) blockers [5].

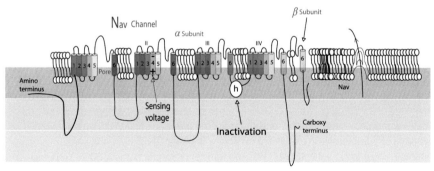

FIGURE 1.8

Primary structures of the α- and β-subunits of the voltage-gated sodium channel [4].

1.4.2 Voltage-gated calcium ion channels

Voltage-gated calcium (Ca^{2+}) channels mediate controlling cellular functions through electrical depolarizing signals. Intracellular Ca^{2+} transients are responsible for many physiological events including neurotransmission, cell cycle, cardiac action potential, muscular contraction, gene expression, and protein modulation. For instance, Ca_V1 channels play an important role in regulating gene transcription [6]. The $Ca_V1.1$ calcium channel contributes to excitation-contraction coupling in skeletal muscle based on the coupling of electrical excitation of the muscle, and so the release of calcium from the sarcoplasmic reticulum (SR) [7]. For this, the $Ca_V1.1$ upon depolarization causes the ryanodine receptors (RyRs), which are intracellular cation channels, to change voluminous release of Ca^{2+} from the SR. Fig. 1.9 shows the subunit architecture of calcium channels. Calcium voltage-gated ion channels are encoded based on the CACNA1S genetic instructions. For instance, the α-subunit forming the channel pore is encoded based on CACNA1S genes.

1.4.3 Voltage-gated potassium channels

Voltage-gated potassium channels are the largest ion channel family in the human genome encoding 40 voltage-gated K^+ channels (K_V), which have diverse physiological functions ranging from repolarization action potentials, setting membrane potential, dictating the duration or frequency of action potential, to modulation of Ca^{2+} signaling and cell volume, to control cellular proliferation and migration [9,10]. Voltage-gated potassium channels are products of 40 genes in 12 subfamilies, and the related K_{Ca} channels are encoded by 8 genes in 4 subfamilies. They conduct K^+ ions in or out of the cell membrane depending on their functionality. Fig. 1.10 shows the schematic view of the structure of voltage-gated K^+ channels.

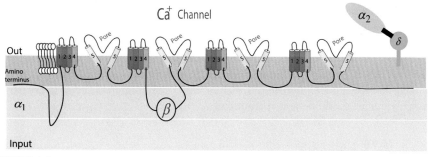

FIGURE 1.9

Schematic representation of the primary subunits of voltage-dependent calcium channels.

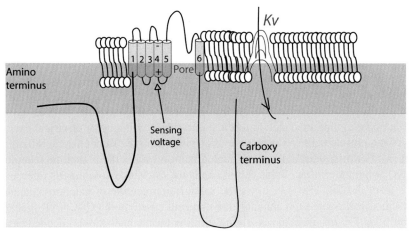

FIGURE 1.10

Schematic representing the primary subunits of voltage-dependent potassium channels.

1.4.4 Inward transient voltage-gated potassium channels

Having diverse physiological functions depending on where they are located, inwardly rectifying K^+ (K_{ir}) channels facilitate potassium ions to move more easily inside the channel. These channels contribute in fixing the resting membrane potential and regulation of electrical excitation. They are not following H-H kinetics (has been explained in Section 1.5). In reverse on what Nernst equation predicts (outward rectification), this channel showed larger potassium ionic flow into rather than out of the cell. There are seven different types of K_{ir} channels that they can be categorized into four functional groups: G protein-gated K_{ir} channels ($K_{ir}3.x$), ATP-sensitive K^+ channels ($K_{ir}6.x$), classical K_{ir} channels ($K_{ir}2.x$), and K^+ transport channels ($K_{ir}1.x$, $K_{ir}4.x$, $K_{ir}5.x$, and $K_{ir}7.x$). This classification is based on their degree of rectification and their functionality with cellular signals.

  Four subunits form the inward rectifiers (K_{ir}), all of which contain two transmembrane segments (M_1 and M_2) like the fifth and sixth transmembrane segments (S5 and S6) of voltage-gated K^+ channels. In addition, a segment with pore elements is located in between. In some cases, this inward current causes the blockade of outward currents by intracellular cations such as Mg^{2+} and through the naturally occurring polyamines spermine and spermidine.

1.4.5 TRPV channels

TRP channels act as signal transducers as they convert sensing parameters such as temperature to changes in membrane potential or intracellular calcium (Ca^{2+}) concentration. After the discovery of a phenotype in *Drosophila* that exhibited as blindness in the presence of constant bright light, the field of TRP channels begun [10]. This superfamily includes six subfamilies: TRPC (Canonical), TRPV

(Vanilloid), TRPM (Melastatin), TRPA (Ankyrin), TRPML (Mucolipin), and TRPP (Polycystic) [11]. As these channels are responding by activation of calcium and sodium fluxes, they can be one target for neural stimulation. One example is photo-thermal stimulation of cells through TRPV channels by using conjugated organic polymers [12].

1.5 Hodgkin-Huxley model

The landmark work of Hodgkin and Huxley describes the permeability of the cell membrane as a function of membrane potential and contribution of different ionic currents in action potential shape quantitatively [13]. This model was based on analysis of voltage-gated channels on the giant axons of squids. It describes the gating mechanism with activation and inactivation of ion channels. Gating particles are a function of time- and voltage-dependent probability functions, which vary between 0 and 1. For instance, based on the H-H model, the Nav channel has three identical activation particles (m^3) and one inactivation particle (h), while experiments provide direct evidence that coupling interactions between voltage sensors in the sodium channel are cooperative [9–11]. Fig. 1.11 illustrates the diagram of H-H model in which V_{clamp} is applied across the cell membrane to measure cell membrane currents, and this configuration is called voltage clamp. I_{clamp} is injected into the cell to measure membrane potential, and this configuration is called current clamp.

The H-H model introduced an empirical kinetic description of ionic conductance, which provides us with practical calculations of electrical responses, and is good enough to predict correctly the action potential shape and conduction velocity. Their model includes mathematical equations that are based on features of the gating mechanisms. Based on the Kirchhoff's circuit laws, sum of the currents flowing through a cell should be equal to the sum of the currents flowing out of the cell.

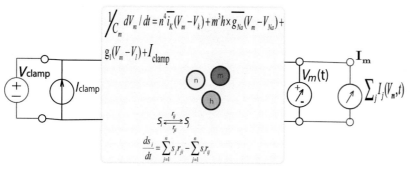

FIGURE 1.11

Shows the diagram of H-H model. V_{clamp} is the input applied voltage across the cell membrane to measure ionic currents, and I_{clamp} is input applied current to monitor action potential or the displacement in the membrane potential.

In Eq. (1.12), H-H model considers that membrane current includes injected stimulus current I_{stim}, capacitive ionic current (I_{C_m}), potassium (I_k), sodium (I_{Na}), and leakage current ($I_{leakage}$). In this model, maximum conductance of g_{Na}, g_K, and g_l are multiplied by coefficients (gating particles) representing the fraction of opening of ion channels.

$$I_m = I_{C_m} + \sum_j I_j + I_{stim} \qquad (1.12)$$

$$I_m = C_m \frac{dV_m}{dt} + I_{Na} + I_k + I_{leakage} \qquad (1.13)$$

$$I_{Na} = \overline{g_{Na}} m^3 h (V_m - V_{Na}) \qquad (1.14)$$

$$I_K = \overline{g_k} n^4 (V_m - V_k) \qquad (1.15)$$

$$I_{leakage} = \overline{g_l}(V_m - V_L) \qquad (1.16)$$

Fig. 1.12 shows the membrane of a single neuron containing the voltage-gated ion channels and the electrical equivalent circuit describing ionic conductances and membrane capacitance based on H-H model. Considering that at resting membrane, the membrane current is initially at zero, based on the electrical equivalence model, if we set the membrane potential at a constant potential by applying a step signal, then we will have

$$-C_m K \delta(t) = \overline{g_K}(U(t) - V_K) + \overline{g_{Na}}(U(t) - V_{Na}) + \overline{g_l}(U(t) - V)_l \qquad (1.17)$$

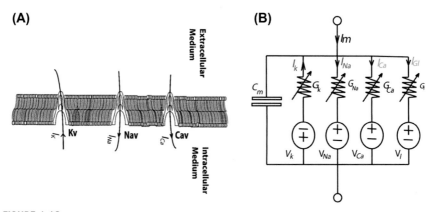

FIGURE 1.12

(A) Schematic of cell membrane including voltage-gated ion channels. (B) Circuit model of neurons developed from H-H model. The cell membrane capacitance (C_m) is originated from lipid bilayer surrounded by the ionic charges from intracellular and extracellular space, the ionic conductance of G_k represents the potassium conductance, the ionic conductance of G_{Na} represents the sodium conductance, G_{Ca} represents the calcium conductance, and G_l is the leakage current.

where the first element of this equation is a capacitive transient current (I_{C_m}), the second ionic current is an outward potassium current (I_k), the third ionic current is inward sodium current (I_{Na}), the last element is the leakage current ($I_{leakage}$), $U(t)$ is step function, and $\delta(t)$ is an impulse signal with the amplitude of K. As impulse signals happen very fast and they are transient currents, we can ignore them in analysis of voltage-gated ionic currents in steady-state conditions.

It is easy to describe g_K at first. Applying a positive step signal across the cell membrane leads to cell depolarization, and g_K increases like an S-shaped time course. On the other hand, during the repolarization phase, the decrease is exponential (see Fig. 1.13). Based on H-H model, several independent gating particles are responsible for such kinetics. To form an S-shaped response, four gating particles should cooperate in an identical manner, each with the probability n to set up an open channel. The probability n^4 occurs when all four particles are correctly placed. As measured K channels are voltage-dependent, the gating particles are supposed to have an electrical charge. Therefore, when the membrane potential is changed, the particle moves between its permissive and nonpermissive positions based on first-order kinetics. Finally, in the steady state, the distribution of gating particles of n relaxes with an exponential time course toward a new value. Fig. 1.13 shows that after depolarization of the cell membrane by a positive pulse, n rises exponentially from zero, and n^4 forms an S-shaped curve, resembling the delayed increasing

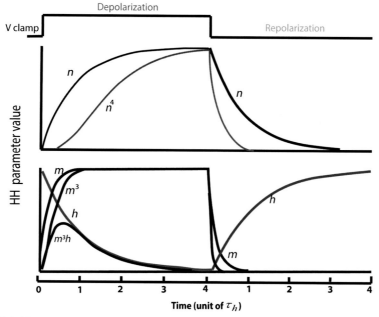

FIGURE 1.13

Time response of gating particles to a positive pulse input.

g_K on depolarization. When the membrane is set again to zero, in repolarization phase, n and n^4 fall exponentially again to zero, which resembles the decrease of potassium conductance.

In Eq. (1.18), illustrating the mathematical form of the time response of potassium gating mechanism, n is a function of voltage- and time-dependent first-order reaction as following:

$$1 - n(Shut) \underset{\beta_n}{\overset{\alpha_n}{\rightleftharpoons}} n(open) \tag{1.18}$$

where α_n and β_n are rate constants for opening and closing of the potassium channels.

$$\frac{dn}{dt} = \alpha_n(1-n) - \beta_n n \tag{1.19}$$

$$\alpha_n = \frac{0.01(10 - v_m)}{\left[\exp\dfrac{10 - v_m}{10} - 1\right]} \tag{1.20}$$

$$\beta_n = 0.125 \, \exp\left(\frac{-v_m}{80}\right) \tag{1.21}$$

We can rewrite the probability equation (Eq. 1.19) based on the voltage-dependent time constants as following:

$$\frac{dn}{dt} = \frac{n_\infty - n}{\tau_n} \tag{1.22}$$

where τ_n is a voltage-dependent time constant and n_∞ is the steady-state value of the probability parameter of n, and they can be obtained from the following equations:

$$\tau_n = \frac{1}{\alpha_n + \beta_n} \tag{1.23}$$

$$n_\infty = \frac{\alpha_n}{\alpha_n + \beta_n} \tag{1.24}$$

Sodium channels as described in Section 1.4.1 have four gating particles. To describe I_{Na} in the H-H model, four hypothetical gating particles were defined, including three activation particles of m and one inactivation particle of h. Eq. (1.14) represents sodium current. Similar to the equation described for the potassium channels, m and h particles are following first-order differential equations. While m rises rapidly and h falls slowly during depolarization, m falls rapidly, and h recovers slowly in the repolarization phase. In addition, m^3h eventually falls to a low value again, which describes increase of sodium current at the beginning of depolarization phase (See Fig. 1.13). Although sodium channels have been found to have identical coupled gating particles, three m and h particles also satisfy the description of the kinetics and the behavior of sodium channels. In addition, in the H-H model, activation and inactivation are independent (at $t = 0$) and depend on the membrane potential.

However, in the main equation they are coupled, as one can affect another by membrane potential. Like the n parameter, m and h also are assumed to follow first-order transition kinetics:

$$1 - m \underset{\beta_m}{\overset{\alpha_m}{\rightleftarrows}} m \tag{1.25}$$

$$1 - h \underset{\beta_h}{\overset{\alpha_h}{\rightleftarrows}} h \tag{1.26}$$

With rates that could be calculated by the first-order differential equations,

$$\frac{dm}{dt} = \alpha_m(1 - m) - \beta_m m = \frac{m_\infty - m}{\tau_m} \tag{1.27}$$

$$\frac{dh}{dt} = \alpha_h(1 - h) - \beta_h h = \frac{h_\infty - h}{\tau_h} \tag{1.28}$$

where

$$\tau_m = \frac{1}{\alpha_m + \beta_m} \tag{1.29}$$

$$\tau_h = \frac{1}{\alpha_h + \beta_h} \tag{1.30}$$

$$m_\infty = \frac{\alpha_m}{\alpha_m + \beta_m} \tag{1.31}$$

$$h_\infty = \frac{\alpha_h}{\alpha_h + \beta_h} \tag{1.32}$$

In the H-H equation, g_L is a background leakage conductance.

Herein, we introduce calcium current, I_{Ca}, which has little contribution in action potential, as it is smaller and slower than sodium current. However, during long depolarization, it dominates the inward current, as it inactivates very slowly. In addition, as a result of intracellular calcium accumulation, the channel is blocked. This shows that inactivation is primarily current-dependent rather than voltage-dependent. The threshold voltage for the activation of calcium channel is about -32 mV and the calcium current could be obtained by the following equations [14]:

$$I_{Ca} = \overline{g_{Ca}} m_{Ca} h_{Ca} \cdot (V_m - V_{Ca}) \tag{1.33}$$

$$h_{Ca} = \frac{K}{K + [Ca^{2+}]_n} \tag{1.34}$$

where the halfway inactivation concentration, K, is a constant and the concentration of free calcium in the shell just below the membrane.

1.6 The Hodgkin-Huxley model predicts action potential shape

The H-H model shows that membrane potential changes can be predicted based on known and measured ionic conductance parameters (Eqs. 1.13−1.16). On the other hand, possible changes in ionic currents can be predicted from the action potential shape, since the contribution and activity of different ionic currents in different phases of action potential has been examined. In addition, this model enables us to calculate subthreshold responses, a sharp threshold for triggering action potential, propagated action potentials in time and space, total membrane impedance changes, and other axonal properties. For instance, Fig. 1.14 shows the diagram of the "dynamic" current clamp system (detailed information about current clamp system has been discussed in Chapter 2), which includes a feedback loop: the patch clamp amplifier measures membrane potential, and calculated sodium ionic current is injected to the cell membrane based on a general computational model of ion channels introduced by H-H model parameters. In the "dynamic" current clamp, a feedback loop is used: the membrane voltage is recorded, and a computational model generates an ionic current based on the recorded membrane potential. This current is then injected into the cell, which in turn changes the membrane potential again, and the whole process is repeated in real time. Afterward, the ordinary differential

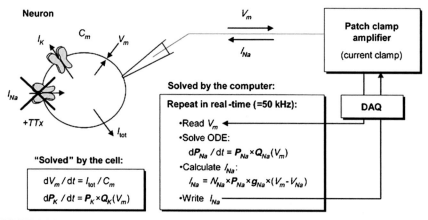

FIGURE 1.14

Finding H-H parameters by using dynamic clamp. As shown, sodium current is blocked with TTX, and instead a simulated current is injected based on a Na_v kinetic model. In the loop, the membrane potential V_m is recorded from the amplifier through the digital acquisition card (DAQ), and the ODEs of the Na_v model are calculated (shown as Markov), and an output current I_{Na} is obtained and injected into the cell. This hybrid biological-computational simulator "solves" the ODEs for V_m, I_K, and other currents with an obtained value of I_{Na} from the kinetic model; at the same time, a new value for I_{Na} is calculated using an effectively constant V_m, as provided by the A/D converter [15].

equations are solved for every time that the current is injected till the point at which the obtained action potential fits with the experimental one (the action potential which is generated naturally without blocking sodium channels) with minimum error.

To find parameters which fit action potential, the followings steps are recommended:

1. Recording action potential by depolarizing the cell membrane and inducing spikes.
2. Blocking sodium channels (e.g., Na_v) and current by using toxins.
3. Based on the H-H model and voltage-dependent parameters, ionic current is calculated and injected into the cell. The parameters are tuned until spiking is restored.
4. The error between the recorded action potential without ion channel blocker (in Step 1) and the dynamic clamp-generated action potentials is calculated.
5. For an error less than a minimum value, the obtained optimal parameters are accepted. If the error is larger than a minimum value, a new set of parameters is taken, and Step 3 is repeated to minimize the error.

1.7 Ionic currents and action potential shape

To show the contribution of different ionic currents in the action potential shape experimentally, time response of action potential recorded in current-clamp modes is correlated to the contribution of ionic currents recorded in voltage-clamp mode while the membrane is scanned in the range of action membrane potential (-70 to $+40$ mV). For instance, in Fig. 1.15, there are various voltage-gated ionic currents in human-induced pluripotent stem cell-derived cardiomyocytes studied by using voltage-clamp technique [13]. According to the figure, in the different phases of action potential, different voltage-gated ion channels are active. The raising phase of the action potential is depending on the fast I_{Na}. After depolarization by inward sodium current, transient outward potassium current (I_{to}) initiates the repolarization phase of action potential. The calcium current (I_{Ca}) is contributing in the plateau phase of action potential. At the starting of falling phase of action potential, mainly rapid rectifier potassium current (I_{Kr}) and slow rectifier potassium current (I_{Ks}) are activated. In addition, inward rectifying potassium current (I_{K1}) is activated during the end of the repolarization phase to stabilize resting membrane potential.

1.8 Summary

Understanding neuronal cell structure, excitable membranes, the superfamily of voltage-gated ion channels, and the landmark work of H-H model is crucial for analysis of neuronal activities based on the patch-clamp and optical electrophysiology.

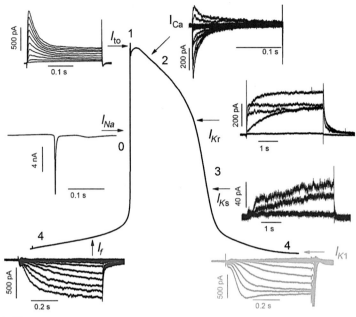

FIGURE 1.15

Shows voltage-gated ionic current traces from hiPSC-CMs and their contribution in different phases of action potential. I_{to}, transient outward potassium current; I_{Ca}, calcium current; I_{Na}, sodium current; I_{Kr}/I_{Ks}, rapid/slow rectifier potassium current; I_f, funny current; I_{K1}, inward rectifier potassium current [16].

Considering the cell membrane as a nonlinear electrical capacitance, charging this capacitance with external ionic currents or through displacement of this capacitance by applying an external electric field induces membrane potential changes. Membrane potential is a key factor controlling neuronal activities. As the membrane hyperpolarization leads to neuronal silencing, membrane depolarization causes neuronal excitation. Moreover, the rule of neuronal interfaces in neuromodulation and neuronal manipulation is based on the alternation of ionic concentrations through voltage-gated ion channels and cell capacitance. Therefore, the connection of the neuronal interface to the cell membrane is important for an effective electrical charge induction (will be discussed in Chapter 2 in detail).

References

[1] Patergnani S, et al. Calcium signaling around mitochondria associated membranes (MAMs). Cell Communication and Signaling 2011;9(1):19.

[2] Frank HY, Catterall WA. The VGL-chanome: a protein superfamily specialized for electrical signaling and ionic homeostasis. Science Signaling 2004;2004(253):re15.

[3] Isacoff EY, Jan LY, Minor Jr DL. Conduits of life's spark: a perspective on ion channel research since the birth of neuron. Neuron 2013;80(3):658−74.

[4] Benarroch EE. Sodium channels and pain. Neurology 2007;68(3):233−6.

[5] de Lera Ruiz M, Kraus RL. Voltage-gated sodium channels: structure, function, pharmacology, and clinical indications. Journal of Medicinal Chemistry 2015;58(18): 7093−118.

[6] Catterall WA. Voltage-gated calcium channels. Cold Spring Harbor Perspectives in Biology 2011;3(8):a003947.

[7] Tuluc P, et al. A $Ca_V1.1$ Ca^{2+} channel splice variant with high conductance and voltage-sensitivity alters EC coupling in developing skeletal muscle. Biophysical Journal 2009; 96(1):35−44.

[8] Wulff H, Castle NA, Pardo LA. Voltage-gated potassium channels as therapeutic targets. Nature Reviews Drug Discovery 2009;8(12):982−1001.

[9] Ranjan R, et al. A kinetic map of the homomeric voltage-gated potassium channel (K_v) family. Frontiers in Cellular Neuroscience 2019;13:358.

[10] Pollock JA, et al. TRP, a protein essential for inositide-mediated Ca^{2+} influx is localized adjacent to the calcium stores in Drosophila photoreceptors. Journal of Neuroscience 1995;15(5):3747−60.

[11] Venkatachalam K, Montell C. TRP channels. Annual Review of Biochemistry 2007;76: 387−417.

[12] Lodola F, et al. Conjugated polymers mediate effective activation of the mammalian ion channel transient receptor potential vanilloid 1. Scientific Reports 2017;7(1):1−10.

[13] Hodgkin AL, Huxley AF. A quantitative description of membrane current and its application to conduction and excitation in nerve. The Journal of Physiology 1952;117(4): 500−44.

[14] Koch C, Segev I. Methods in neuronal modeling: from ions to networks. MIT Press; 1998.

[15] Milescu LS, et al. Real-time kinetic modeling of voltage-gated ion channels using dynamic clamp. Biophysical Journal 2008;95(1):66−87.

[16] Prajapati C, Pölönen R-P, Aalto-Setälä K. Simultaneous recordings of action potentials and calcium transients from human induced pluripotent stem cell derived cardiomyocytes. Biology Open 2018;7(7):bio035030.

Principle of whole-cell patch-clamp and its applications in neural interface studies

2.1 Introduction

Advancement of new technologies in neuroprosthetic devices and pharmacological developments for curing cellular diseases demands studying of electrophysiological behaviors of cells to evaluate the efficacy of treatments based on patch-clamp experiments. In fact, in primary in vitro studies, basic voltage- and current-clamp experiments for interested grown cells with voltage-gated ion channels are performed to measure membrane potential and ionic conductances. Studying complicated behaviors of the different ion channels in different types of cells needs high accuracy of the measurements and mathematical modeling from single whole-cell recordings. For instance, recently emerging organic photoactive polymers in solar cells have been employed in retinal cell stimulation for curing blindness in cases of photoreceptor regenerations [1]. In some conditions, photostimulation of neurons mediated with these polymers could result in hyperpolarization (long light pulses), while in other cases, it could result in cell depolarization (short light pulses). Moreover in some cases, rebound or anode break stimulation can stimulate action potential by hyperpolarization of a cell [2]. In others, local heat can modulate membrane capacitance, leading to action potential generations [3]. Discovery of all these mechanisms has been facilitated by patch-clamp technology.

In this chapter, first the whole setup and different components of a patch-clamp system have been explained as a general introduction. Then, to start with whole-cell recording, charge transfer mechanisms in an electrolyte solution medium has been explained. Afterward, two types of charge transfer mechanisms of Faradaic and capacitive have been described. Then protocols for recording different ionic currents by patch-clamp technique from a cell have been introduced in a practical manner. Moreover, in this chapter because of the importance of the landmark work of H-H (introduced a quantitative description and an electrical model of a giant nerve fiber), it was employed to illustrate that how we can extract kinetics of ion channels. Therefore, extraction of H-H parameters for understanding neurophysiological behaviors of cells based on patch-clamp studies has been mentioned.

Electrophysiology Measurements for Studying Neural Interfaces. https://doi.org/10.1016/B978-0-12-817070-0.00002-6

Furthermore, in this chapter, two different types of Faradaic and capacitive neural stimulations are discussed to classify required considerations for each case in patch-clamp studies. As Faradaic stimulation uses electrochemical ionic current to change membrane potential, membrane resistance will be affected during the stimulation; however, in capacitive stimulation, membrane capacitance is changed through an external electrical or thermal stimulus. Moreover, this chapter also covers different open- or closed-loop biointerface structures for achieving highest efficiency and safety of neural stimulations. In all studies, whole-cell recording played a significant role in understanding the performance of the biointerfaces. For example, in different neural interfaces, such as photoelectrodes made of semiconductor quantum dots (QDs), Faradaic currents are generated. These can depolarize or hyperpolarize cells under light illumination. On the other hand, in metallic nanoparticles, heating the membrane could lead to membrane capacitance changes and cause depolarization for short illumination, while in some conditions, it can hyperpolarize cells under long illuminations.

2.2 Electrophysiology setup

Fig. 2.1 shows a general schematic of an electrophysiology setup. Herein, we introduce generally different components of this setup and steps to patch a cell and record the electrophysiological properties of cells. At the beginning, cultured cells (grown on a substrate) are transferred in ionic solution of artificial cerebrospinal fluid (aCSF) medium proper for electrophysiology measurements. To apply an ionic current through chloride ions, a silver/silver chloride (Ag/AgCl) electrode is employed for a reversible electrochemical reaction and to create ionic current without corrosion. In addition, an extra Ag/AgCl electrode is used for collecting electrical ionic charges and grounding the extracellular medium. Cultured cells in aCSF medium are found and observed by the optical microscope setup (objective, light filter, light polarizers, etc.). A nanopositioner is used to approach and patch a cell. The electrical signals are applied with a signal generator in data acquisition (D/A) unit, the electrical properties of the cells' membrane are measured with (D/A) unit, and a highly accurate electrophysiology amplifier (the electronic circuits and functions of which have been explained in Chapter 6) amplifies electrical signals from the cells and D/A unit (in case the input signal needs amplification).

2.3 Capillary glass electrodes

Capillary glasses are made with a shutter instrument in a pulling process to make a fine pipette tip. A very small pipette tip (about 1um) is produced for measuring very low amounts of ionic currents (picoamps), and it also allows patching just a small part of the cell membrane. In this way, Ag/AgCl electrode will be immersed in a

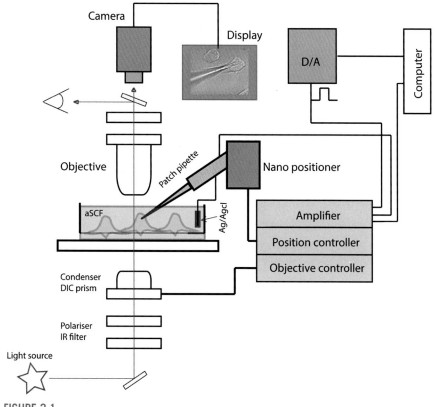

FIGURE 2.1

Schematic of an electrophysiology setup. It includes optical components such as objective, magnification, light filter, and a light polarizer to observe the cells. A data acquisition unit (D/A) is used to generate favorite signals and convert output analog signals to digital. To monitor the cells, a digital camera and display are employed. An amplifier also boosts small signals recorded from the cells and from the D/A which generates favorite signals. A nanopositioner facilitates approaching cells. A positioner and objective controller are used to find the cells in a sample plate. A patch pipette including an Ag/AgCl electrode is used to patch the cells. A ground Ag/AgCl electrode is used to collect ionic charges and make a closed-loop electrical circuit with the Ag/AgCl electrode immersed in the patch pipette filled by intracellular medium.

filled capillary glass pipette to connect inside a cell membrane electrically. There is one pressure tube connecting to the tip for controlling pressure (for applying positive and negative pressures) to get a good giga-ohm seal. Fig. 2.2A shows a cell-attached patch and the schematic of a capillary glass electrode filled with intracellular medium. Approaching and connecting the pipette tip with the cell membrane has many steps; it can be monitored with two optical visualizations (with optical microscope) as well as electrical measurement of the patch pipette's resistance (see

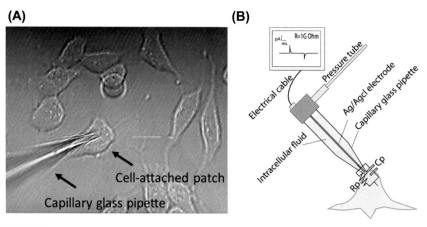

FIGURE 2.2

In vitro single-cell patch clamp. (A) Patching an SHSY-5Y cell; (B) patch pipette's head consists of an Ag/AgCl electrode, pulled capillary glass, intracellular fluid, and a pressure tube.

Fig. 2.2B and 2.3A). During the process of patching a cell, with applying electrical voltage into the tip, the tip's resistance can be monitored online. This enables tracking resistance changes of the pipette's tip during the patching process. Therefore, while patching a cell membrane, measured electrical resistance measure with pulse test (see Fig. 2.3B) should be increased and required negative pressure applied to get a giga-ohm seal afterward. After attaching the pipette tip with the cell membrane, the resistance of the tip increases as the tip will be surrounded by the membrane. Fig. 2.3D−G illustrates the process of attachment for the tip with the surface of a cell membrane till its resistance goes up to giga ohm.

2.4 Measurement principle

aCSF is a biological solution medium that enables us to measure electrical properties of neuron cell membranes such as membrane potential, conductance, and capacitance while the cells survive in conditions similar to those of the brain. Electrical connection of a neuron cell in aCSF medium with the Ag/AgCl electrodes is similar to a chemical cell. In Fig. 2.4, this electrochemical cell consists of anode and cathode electrodes, both of which are Ag/AgCl. However, one is immersed in intracellular and another one in extracellular medium, while the cell membrane is separating the intracellular and extracellular medium, voltage-gated proteins are defining the electrical ionic conductance. The electrochemical properties of this electrochemical cell depend on the resting membrane potential and ionic membrane conductances. Fig. 2.4 shows the electrochemical cell configuration of a cell with the electrodes, in which one (intracellular Ag/AgCl) is for

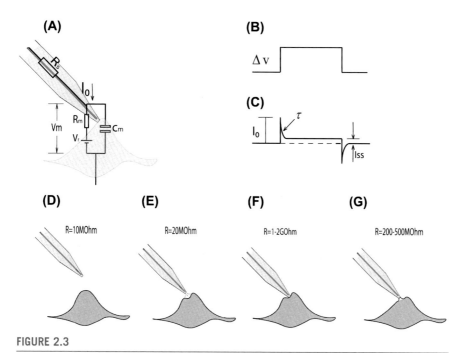

FIGURE 2.3

Schematic of patching process of a cell. (A) Patch pipette tip is far from the membrane, and the tip resistance is 10 MΩ. (B,C) applied pulse voltage to the cell membrane and recorded current. (D) After getting close to the membrane, positive pressure through pressure tube is applied to push cell membrane and reach the surface to form a fine seal. (E,F) After the resistance of the tip reaches a significant increase, continuous suction leads to a giga-ohm resistance as the tip is totally surrounded by the membrane. (G) In the final step, suction can lead to whole seal and the resistance decreases to the membrane resistance, which is around several hundred Mega Ohm.

injecting electrical charges and inducing negative or positive potentials, while the second Ag/AgCl is grounding the extracellular medium and collecting ionic charges. Upon applying a positive potential with a function generator (V_h), electrons will leave the intracellular electrode, and positive ionic charges will remain inside the cell, which depolarizes intracellular medium. On the other hand, the same amount of electrons is grounded at the second Ag/AgCl electrode, leading to a closed-loop circuit. As the figure shows, the intracellular electrode has a resistance and capacitance as a result of the capillary glass tip and the connection of the electrode with the inside of the membrane, and the second part is the equivalent electrical circuit of a cell in resting state, including membrane resistances, capacitance, and resting potential. The membrane conductance is altered by the change in permeability of the voltage-gated ion channels, and the membrane capacitance depends on the cell geometry and charge distribution around the lipid bilayer, and resting membrane is generally around −60mv for a neuron cell.

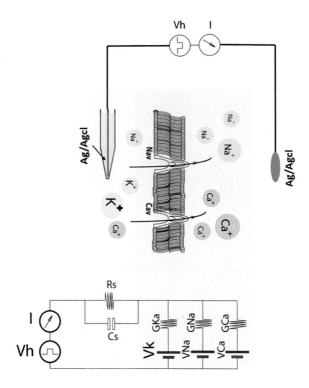

FIGURE 2.4

Schematic of a patched neuron cell under whole-cell measurement resembling an electrochemical cell, in which Ag/AgCl electrodes are anode/cathode electrodes and neuron cell including resting membrane potential and its resistances are in membrane layer.

To measure the conductance properties of a neuron (current/voltage = I/V), different voltages are applied by the voltage generator $V\underline{s}$, and the membrane current is measured by a picoammeter A (in Fig. 2.4). While the voltage is applied, based on the voltage level, some ion channels get activated and so the membrane current and its conductance are changed.

2.5 Charge transfer mechanism

To depolarize and hyperpolarize neuron cells, injected electrical charges from the signal generator should be able to ionize the medium in a positive and negative manner, respectively. For that, microelectrodes should work in two directions and electrons could be transferred from the aCSF solution to the microelectrodes and vice versa. Moreover, ionizing the aCSF medium by microelectrodes should not cause any corrosion or damage to the microelectrodes by chemical reactions.

Therefore, different materials and interfaces are employed for an efficient and safe charge transfer. Fig. 2.5 shows the schematic of different microelectrodes and the charge transfer mechanism for each one. For Ag/AgCl microelectrodes, chloride ions are charge carriers for ionization. The Ag/AgCl electrode can give an electron to the electrolyte medium by sending a Cl^- ion with Ag/AgCl dissociation to $Ag + Cl^-$ while the remained Ag atom can be taken by the silver wire (see Fig. 2.5A). The Ag/AgCl coating can reduce an electron from the solution by absorption of a Cl^- ion and taking a silver atom (Ag) from the silver wire. Fig. 2.5B shows charge transfer mechanism between the platinum electrode (Pt) and aCSF ionic solution that is based on the hydrogen adsorption-desorption reaction acting like a capacitor. This hydrogen adsorption reaction has been described by Eq. (2.1). In which H^+ is a proton, Pt represents a specific site on the platinum surface, and e^- is an electron transferred from the electrode to the solution upon applying a negative potential to the electrode. Eq. (2.2) represents the hydrogen desorption, in which upon applying a positive potential to the electrode, the electron transfers back to the metal and the proton returns back to the solution:

$$Pt + H_2O \leftrightarrow PtO + 2H^+ + 2e^- \tag{2.1}$$

$$Pt + H_2O + e^- \leftrightarrow Pt - H + OH^- \tag{2.2}$$

2.5.1 Faradaic and capacitive charge transfer

Understanding the charge transfer mechanism from electrodes to aCSF medium is crucial in studying neural stimulation. There are two different mechanisms of capacitive (non-Faradaic) and Faradaic charge transfer at electrode/electrolyte interface.

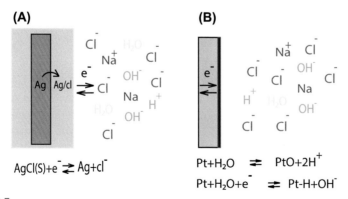

FIGURE 2.5

Schematic of charge transfer for Ag/AgCl and platinum electrodes in aCSF solution. (A) Charge transfer from Ag/AgCl electrode and electrolyte; (B) charge transfer mechanism for platinum electrode/electrolyte interface.

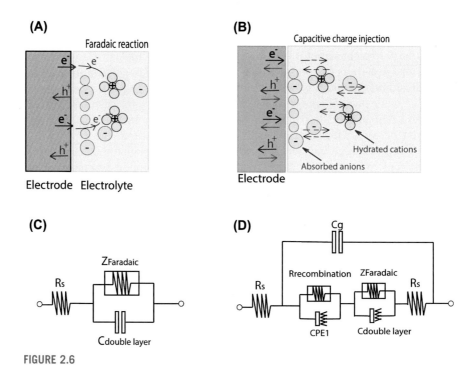

FIGURE 2.6

Schematic of electrode/electrolyte interface for both Faradaic and capacitive charge transfer. (A) Faradaic interface in which the electrochemical irreversible reaction at the interface of the electrolyte medium is the dominant mechanism for charge transfer; (B) capacitive interface in which charge accumulation and movement at the interface induces reversible redistribution of charges at the medium. (C) Electrical circuit model for mechanism of charge transfer of an interface with dominant Faradaic behavior. (D) Electrical circuit model for a multilayer electrode with dominant capacitive behavior.

As Fig. 2.6A and B illustrates, in Faradaic mechanism, electrons are transferred to the solution, leading to oxidation/reduction of the chemical species at the electrode/electrolyte interface; however, in capacitive (non-Faradaic) mechanism, redistribution of charged ions without electron transfer from electrode to the solution medium occurs. Fig. 2.6C shows a simple electrical model for a Faradaic interface in which C_{dl} is the double layer as a result of several physical phenomena [4]. The first reason for the creation of this double layer is because when a metal is placed in electrolyte, charge redistribution occurs as electrons at the surface react with the electrolyte and a transient electron transfer leads to a depletion layer at the interface or a planar charge layer. The second reason is that polar molecules like water have different orientations at the interface, and a net orientation causes charge separation at the end. The Faradaic resistance in Fig. 2.6C appears when net charge on metal is forced to vary due to external stimulation, which leads redistribution of charges in the solution. The Faradaic process includes transfer of electrons from the metal electrode and reducing hydrated cations in the electrolyte. In addition, the direction of charge transfer is controlled by the direction of external electric field applied to the electrode, and the electron transfer may be from the electrode to the aCSF electrolyte or reverse (oxidation/reduction).

Fig. 2.6D shows an electrical model for a complex capacitive electrode in which the capacitive charge transfer is dominant. In this circuit, R_s is the resistance due to electrode connection with electrical probe, C_g is the characteristic capacitance of the electrode, R_{rec} is the recombination process of charge carriers, ZF is Faradaic resistance at the electrode-electrolyte interface, CPE2 represents the capacitive nature of the electrode-electrolyte interface which is parallel to the ZF, and RL is the leakage resistance of electrolyte media. For instance, in this structure, C_g of the electrode should be high to enhance capacitive behavior.

2.6 General protocol to patch cells

Patch clamp is not a simple technique. It requires many steps, from approaching a cell till the recording step, and there are many details that one should consider at each step. To patch a single cell, the following steps are recommended:

1. A water immersion objective is placed inside the aCSF medium by objective controller to find grown cells: for this step, one needs to search and change the objective positions to observe the cells.
2. Taking the objective away to make space for immersing the patch pipette's tip near the interest area: for this step, one needs to take the objective up and immerse the patch pipette inside the solution. While immersing the pipette inside the aCSF solution, one can blow inside the tube (increase the pressure) to exert positive pressure to prevent any blockage of the patch pipette tip during the immersion in aCSF medium.
3. Alignment of the objective on top of the patch tip inside the solution: to do this, we have to immerse the objective inside the solution in a way that enables us to observe the patch tip without touching and breaking it. It is a critical step, as it requires high accuracy and precision to find and observe the immersed patch tip inside the solution as the tip size is around 1um. A good trick is to immerse the objective in an area adjacent to the patch tip and try to align the objective with a higher safe Z position.
4. Removing offset potential and reading pipette's resistance immersed inside the aCSF: after putting the patch pipette tip inside the solution, offset potential should be reduced to zero by offset adjustment circuit (as explained in Chapter 6).
5. Tracking patch pipette tip with objective: in this step, one needs to bring the patch pipette tip close and nearby cells at the bottom of the culture plate. So that, upon changing the Z position of the patch pipette tip and immersing more in the aCSF solution, the objective's focus and position should be controlled to follow

the patch pipette tip. To prevent touching of the bottom by the patch pipette tip, one needs to first change the objective focus (Z position) to observe deeper positions beforehand changing position of patch pipette tip.

6. Patching a single cell and rupturing of cell membrane: after reaching at the bottom of the plate while following the patch pipette tip, cells can be observed while the patch pipette tip is focused. During touching of cell membrane by the patch tip, applying a positive pressure by blowing inside the tube leads a better patch and breakdown of the membrane to get a giga-ohm seal. After pushing the membrane under patching process, resistance of the forming seal between membrane and the tip of the pulled electrode starts to increase to several 10 MΩ; after reaching a value around 20–40MΩ, a gradual suction (negative pressure inside the tube) can lead to a giga-ohm seal and then the resistance increases, up to several hundred Mega ohm or several giga-ohm. In this step, keeping suction to achieve the giga-ohm seal is necessary.

7. Getting whole seal: in the next step, further suction leads to get the whole cell as it has been illustrated already in Fig. 2.3.

2.7 Voltage clamp

In voltage-clamp mode, a voltage is applied across cell membrane and resulted ionic currents are recorded. Therefore, conductance of the membrane for different membrane potentials can be measured for different voltage scans. Fig. 2.7 shows schematic of a circuit for voltage-clamp mode measurements. In this circuit, a pulse-step voltage is applied at a constant voltage level, and membrane current or conductance is measured through an operational amplifier (the function

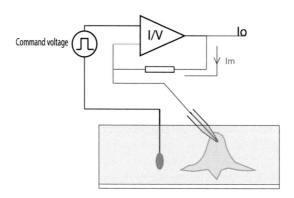

FIGURE 2.7

Schematic of voltage-clamp mode configuration. Measurement setup consists of an amplifier circuit for measuring low amount of ionic currents and a signal generator for applying different voltages across the cell membrane.

of this amplifier has been explained in Chapter 6). Therefore, when the voltage-gated ion channels are opened, conductance of the membrane and transconductance gain of the amplifier are changed, and so the output signal's amplitude is correlated to the membrane's conductance (as explained in detail in Chapter 6).

2.8 Current clamp

In current-clamp mode, a current is injected into the cell membrane while membrane potential is recorded. Generally, to record action potentials or other membrane parameters from membrane potential, we use current-clamp mode. Fig. 2.8 shows schematic of current-clamp mode circuit configuration. In this circuit, a stimulus current signal (red line in the figure shows the injected current path to reach membrane) is injected into the cell membrane, while membrane potential and current are recorded through high-input impedance transconductance amplifiers (as explained in the Chapter 6).

2.9 Whole-cell recordings

To do whole-cell recording, following protocol is recommended:

1. Set a proper holding potential (normally a negative potential equal to a resting potential of a neuron around −60mv is applied).
2. Capacitance compensations (the compensation circuit has been explained in Chapter 6).

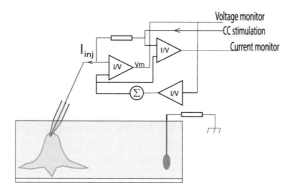

FIGURE 2.8

Schematic of current-clamp mode configuration. Measurement setup consists of an amplifier circuit for measuring low amount of ionic currents and a signal generator for injecting different currents inside the cell membrane through CC stimulation terminal.

3. Voltage pulse scan is used in voltage-clamp mode to measure I-V characteristic of a patched cell (resting membrane potential and resistance of the cell after rupturing the cell membrane).
4. Switch to current-clamp mode to measure membrane potential.

2.10 Extraction of H-H model parameters

To measure electrical properties of the cells for pharmacological applications or evaluating efficacy of neural stimulation, it is necessary to obtain H-H model parameters. In fact, as the relationship of the membrane potential with ionic conductances and membrane's capacitance is complicated (nonlinear), in most studies, H-H parameters are extracted by patch-clamp protocols. To find H-H model parameters for an excitable cell, we need to replicate Hodgkin and Huxley's voltage- and current-clamp experiments, which they did for a giant nerve fiber. For this purpose, we follow all steps of the landmark work of Hodgkin and Huxley: (**1**) finding the time response of potassium, sodium, and calcium ionic conductances based on voltage-clamp experiments (I-V), (**2**) finding how ionic currents vary with membrane potential (calculating activation and inactivation for ionic channels), and (**3**) fitting the responses with formulated differential equations that relate ionic conductances to membrane potential. Finally, with fitting action potential, we can rebuild a new H-H model for the interested cell.

We start from the most important elements of the H-H model, which are ionic conductances. To find the behavior of each ionic conductance separately, we have to know how we can activate and inactivate them at first, as follows:

1. Applying a positive potential across membrane leads to depolarization of the cell membrane, which activates Na^+ and Ca^+ channels: upon membrane depolarization, sodium and calcium ions enter cell membrane.
2. Applying a negative potential across membrane leads to hyperpolarization of the cell membrane and K^+ channels get activated: upon membrane hyperpolarization, potassium ions enter cell membrane (potassium VGCs are active, sodium and calcium VGCs are inactive).
3. To measure Na^+ ionic currents separately, we can use K^+ channel blockers such as TEA. In this case, sodium ionic current and Nernst potential can be obtained from voltage-clamp experiments.
4. To measure K^+ ionic currents, we can use Na^+ channel blockers such as TTX. In this way, potassium ionic current and Nernst potential can be calculated from voltage-clamp experiments.

2.10.1 Time response of ionic currents (voltage clamp)

Time responses of ionic currents for different applied voltages across cell membrane can be achieved in voltage-clamp experiments. They are a function of both time and membrane potential, and they can be calculated based on Ohm's law as follows:

$$I(t) = g(t, v).(V - E) \tag{2.3}$$

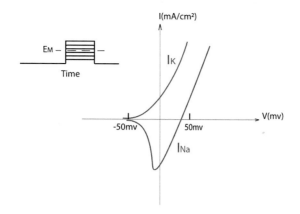

FIGURE 2.9

Current-voltage (I-V) relations of squid axon. Sweep of step voltages is applied across the membrane, and its conductance is recorded. Two measured outward ionic current (blue trace) of I_K and I_{Na} (red trace) are plotted against test potential. The plot shows the opening of both channels between −50 and −20mv.

where I is the recorded membrane current, V is the applied potential across cell membrane, E is the Nernst potential, and g(t, v) is the conductance of ion channel. If we separate ionic currents to two I_{Na} and I_k, Hodgkin and Huxley's next step is finding quantitative measurement of ionic conductances. Fig. 2.9 illustrates the peak of ionic currents of I_{Na} and I_K for different input membrane potentials. In fact, conductance of a cell depends on opening probability of the channels, number of ion channels in the cell membrane, and can be obtained as follows:

$$g(t, v) = N_{total} \cdot y_1 \times P(v, t) = \overline{g_{max}} \times P(v, t) \qquad (2.4)$$

where N_{total} is the total number of ion channels, y_1 is the conductance of one ion channel, and $N_{total} \cdot y_1$ corresponds to the maximum conductance of the total channels when all of them are open, and it is called $\overline{g_{max}}$, and it can be obtained from the peak currents (see Fig. 2.9) over input scan voltage.

Potassium ion conductance: Since potassium (K^+) ions have negative reversal potential, they tend to dampen excitation of neurons. Their diversity is large, and their existence facilitates encoding of action potential firing patterns and rhythmic activity. Basically, K^+ channels stabilize the membrane potential, as they are more negative and keep membrane potential further from the firing threshold. In this way, they adjust resting membrane potential, define the time interval between repetitive action potentials, and generally reduce the effectiveness of input excitation when these channels are activated already. There are different types of K^+ channels that have been reviewed by Adams et al., Moczydlowski et al., and CG Nichols et al. [5−8]. In most excitable cells, high K^+ permeability after action potential generation comes from rapidly activated delayed rectifier K^+ channels. In some encoding membranes, in subthreshold range of membrane potential, fast transient current from this channel is called I_A rapidly inactivity current or transient outward current I_o.

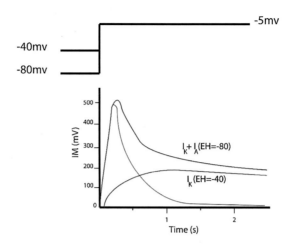

FIGURE 2.10

Time response of potassium current to a −25mv applied voltage pulse across cell membrane.

Activation of K_A channels could occur by a depolarization after a rapid hyper-polarization. Fig. 2.10 shows that I_A could be found and separated from the outward current by an in-out voltage-clamp step with a holding potential. If we choose −40mV holding potential, a step depolarization to −5mv leads to stimulation of K^+ current in delayed rectifier K^+ channels and in Ca^+-dependent K^+ channel, while not in K_A channels. Instead, if we choose −80mv holding potential, the depolarization leads to faster current, transient I_A as well. Subtraction of these two traces results in rapidly activating and inactivating time course of I_A alone (see Fig. 2.10).

Hodgkin and Huxley described the kinetics of I_A by empirical models of $\overline{gA}a^4$, where a^4 gives activation with a sigmoidal raised. The time response of the conductance consists of rising and falling parts. The rising response can be fitted by the $\overline{g_{max}}(1 - e^{-t})^4$, while the falling part by $\overline{g_{max}}(e^{-4t})$; $\overline{g_{max}}$ is the steady state response of the conductance when the conductance is maximum and the ion channels are completely open.

And finally, the potassium conductance can be described as product of the steady-state response of the potassium conductance and the probability function of n:

$$G_k = \overline{g_{max}} \times n^4 \qquad (2.5)$$

Sodium ion conductance: There is less diversity in terms of functionality among Na^+ channels in excitable cells, although they are obviously not all the same. There are significant kinetic differences between fast TTx-sensitive Na^+ and slower TTx-insensitive channels [9]. In axons, Na^+ channels are responsible for generation of the rapid upstroke of the action potential. While potassium channels have only activation probability particles, sodium channels are more complicated as they have voltage-gated activation (m) and voltage-gated inactivation (h) probability particles as well. Their sodium conductance can be found as follows:

$$G_{Na} = \overline{g_{max}} \times m^3 h \tag{2.6}$$

To obtain activation function, we need to sweep the membrane potential based on a protocol to open channels. As shown in Fig. 2.11A, holding membrane potential is adjusted at -95mV and voltage-clamp pulses range from -65mv up to $+5$mv. Fig. 2.11B and C shows obtained current-voltage relationship and normalized ionic conductance (activation) from the sweep. At -95mv (holding potential), sodium channels are not active, while by increasing to -65mv, some of them become activated, and at $+5$mv, most of them are already open.

To obtain inactivation of the channels (h), we hold a cell at a potential at which channels are open already. By applying lower potentials than its Nernst potential, we close them. In this way, we can measure inactivation probability function (h). Fig. 2.11D and E shows inactivation protocol and resulting ionic currents to

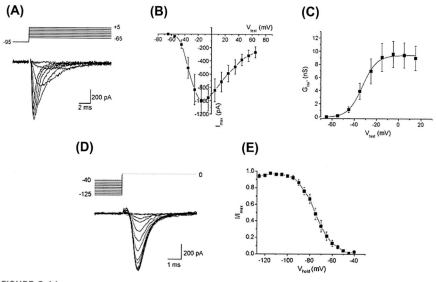

FIGURE 2.11

Activation and inactivation of sodium channels in SJ-RH30 cells. (A) Recorded Na$^+$ ion currents in response to applied sweeps from a typical SJ-RH30 cell. (B) A plot current of current-voltage data from several recordings (C) A conductance-voltage plot from several recordings. (D) Inactivation of Na$^+$ currents upon sweeping membrane potential (-120mv to -40mv). (E) Normalized steady-state data of several recordings like shown in (D) [10].

calculate inactivation for the Na^+ channels. In general, inactivation of Na channels (>95%) occurs by depolarization of the cell membrane to 0 mV and beyond, based on H-H model. As shown, the potential pulses are from −125mv up to −40 mv, while holding potential is set to 0v. It means that the ionic conductance at 0v is high, but applying a negative potential leads to inactivation of the channel. By this protocol, we can find inactivation of the Na^+ channel.

2.10.2 Time response of membrane potential (current clamp)

After obtaining ionic conductances in voltage clamp, membrane potential can be measured in current-clamp mode while an external current can be injected through intracellular electrode into the cell for measuring depolarization and hyperpolarization levels. As mentioned in Chapter 1, membrane potential is a product of ionic currents and conductances. The cell membrane's behavior before a threshold value (depending on the cell type, it is normally around −45mv for neurons) is passive, so by injecting current, we can measure depolarization and hyperpolarization level of the cell membrane (depending on whether the injected ionic current is positive or negative).

2.10.3 Membrane capacitance and resistance calculations

To measure membrane impedance (resistance and capacitance), there are two approaches of electrical pulse response and electrical impedance scanning. Here, we explain the pulse response analysis when the cell is in passive mode. Fig. 2.12 shows the electrical pulse response. As illustrated, with applying a pulse voltage, membrane current and therefore resistance can be measured and tracked during the patch-clamp process. Fig. 2.12A and B shows a pulse voltage (U(t))

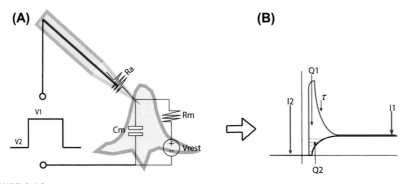

FIGURE 2.12

Pulse response for membrane capacitance and resistance calculation. (A) Electrical model of cell membrane in passive mode and in resting condition; (B) time response of the membrane current to a pulse signal.

applied to the cell membrane, and the resulting capacitive and resistive responses, respectively. Eq. (2.7) describes the pulse response, which includes a transient (capacitive) term and a steady-state (resistive) response in which R_a is access resistance as a result of patch pipette connection with the cell membrane and R_m, C_m, and V_{rest} are the membrane resistance, capacitance, and resting membrane potential, respectively. As membrane capacitance is charging through R_a, time constant of the membrane capacitance is equal to $R_a C_m$, and the maximum transient current peak is equal to V_1/R_a. Therefore, based on Eq. (2.8), R_a, membrane capacitance can be obtained from the transient response, and for the steady-state condition, R_m can be calculated from the second term.

$$I_m = C_m \frac{dv_m}{dt} + \frac{v_m - v_{rest}}{R_m} \tag{2.7}$$

$$I_m = \frac{U(t) - vm}{R_a}, I_{c(\max)} = \frac{V1}{R_a} \tag{2.8}$$

$$U(t) = R_a C_m \underbrace{\frac{dv_m}{dt}}_{\substack{\tau \\ Transient}} + \underbrace{\frac{R_m + 2R_a}{R_m + R_a} v_m - \frac{R_a}{R_m + R_a} v_{rest}}_{Steadystate} \tag{2.9}$$

2.11 Faradaic stimulation of neurons

Based on the discussion in 2.5.1, Faradaic charge transfer could be reversible if a proper electrode such as platinum or Ag/AgCl is used. However, a proper electrical signal should be applied, which leads to charge balance and prevents electrochemical corrosion of the electrode because of charge accumulations [11]. By reduction (adding electrons to the solution medium) and oxidation (taking electrons from the solution) process, the charge may transfer from the electrode to the aCSF electrolyte medium in Faradaic current. Faradaic charge transfer can be categorized into two reversible and irreversible types. The reversibility depends on the relative rates of kinetics (electron transfer at the interface) and mass transport. If the kinetics are fast in comparison with mass transport, then the Faradaic reaction is reversible [12].

2.12 Capacitive stimulation of neurons

A neuron cell can be considered a leaky capacitor. Intracellular and extracellular ionic charges around the lipid bilayer build an electrical capacitance. Fig. 2.13A shows a cell as an enclosed area consisting of an intracellular medium which has less positive charge in comparison with extracellular medium and related produced

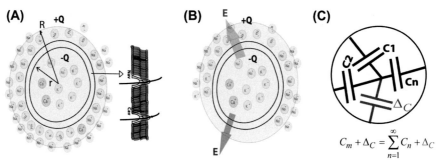

FIGURE 2.13

Schematic of a cell similar to an electrical capacitance. (A) Intracellular charges are separated from the extracellular medium by a lipid bilayer as a dielectric and it creates a capacitance; (B) a disturbance in charge distribution causes a change in electric field and a different electric field distribution at the left bottom side of the cell membrane; (C) the change of the electric field at the bottom-left corner leading to a change in the capacitance per area.

electric field (see Fig. 2.13B), which produces a negative resting potential. Based on Gauss' law, the membrane capacitance can be calculated from Eq. (2.10). As the capacitance directly depends on the charge distribution around the cell membrane, any disturbance or rearrangement that can change membrane capacitance and so membrane potential is called capacitive stimulation. We can consider a cell as many parallel capacitances per area as shown in Fig. 2.13C.

One method for capacitive stimulation is inducing an external electric field across the cell membrane to rearrange ionic charges around lipid bilayer. The second method is injecting a capacitive current into the cell membrane. In this approach, a proper capacitive electrode is needed for producing such a current. The third method is inducing surface potential or charges by using conductive polymers.

For example, consider a cell with $C_m = 76$ pF in resting potential at 60 mV, which has new induced 5 pC charges around its membrane. This much charge can induce extra 76 pF capacitance as a result of newly induced charges across the cell membrane. To null the induced charges, rearrangement of charges across the membrane induces extra membrane potential of 33 mV. This change in membrane potential is enough to excite a neuron cell, as threshold potential for firing an action potential is at about 45 mV, which is 20 mV more than resting potential.

$$\Delta_C = \frac{Q}{V} = \frac{5pC}{60mv} = 76pF \tag{2.10}$$

$$\Delta_V = \frac{Q}{C + \Delta_C} = \frac{5pC}{152pF} \approx 33mV \tag{2.11}$$

To induce an external electrical field for capacitive stimulation and minimum Faradaic current, high impedance and dielectric metal/insulator capacitor structures like metal/hafnium oxide (HfO_2) are employed. In the following, we discuss external stimulation of neurons with metal/HfO_2 capacitive structure and related electrophysiology measurement for neural stimulation experiments.

2.12.1 Silicon capacitors for capacitive stimulation

Extracellular stimulation as a classical approach is widely used for stimulation of cultured brain tissue for neuroprosthetics. By using voltage-clamp and current-clamp measurements of an individual neuron on a semiconductor-insulator interface, a noninvasive capacitive stimulation can be studied. Ingmar Schoen et al. showed capacitive stimulation of pedal ganglia of *Lymnaea stagnalis* by applying voltage ramps to an HfO_2/silicon capacitor [13]. Considering a neuron cell grown on top of a dielectric as shown in Fig. 2.14, with applying a ramp electric potential to the substrate and aCSF solution, there will be an electric field distribution in the medium and across the cell membrane. Upon applying an external voltage to the metal/HfO_2 capacitor, there will be a nonhomogeneous electric field distribution on the cell membrane and membrane-capacitor junction. Herein, the effect of the attached part of a grown cell on top of the interface and the free membrane part is discussed based on two-domain stimulation model. With current-clamp technique, the mechanism of neural stimulation was described for rising voltage ramps, which depolarize free membrane, while falling voltage ramp depolarizes the attached membrane and it leads to firing of action potentials.

In principle, a gradient electrical current in the aCSF medium (cell medium) can result in extracellular potential that causes firing of action potentials. Although possible depolarizing and hyperpolarizing effects on different parts of neurons are explored, extracellular stimulation of culture neurons on metallic substrates is useful to address which part of the cell membrane is responsible for excitation. As in

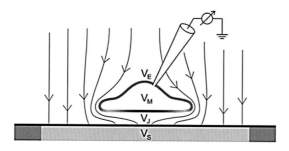

FIGURE 2.14

Capacitive stimulation of a cell membrane on top of a capacitor through applying an external voltage stimulation [14].

extracellular excitation, where a closed-loop current including anode and cathode currents contributes to the stimulation, the role of each current in the cell excitation should be explained. Furthermore, the contribution of Faradaic and capacitive currents in the stimulation is important to distinguish the safety and effectiveness of the stimulation. And finally, we need to understand which form of input current and voltages is better to control stimulation. To answer these questions, Ingmar et al. studied patch-clamp experiments under the following conditions [13]:

1. An insulated planner electrode was used without any Faradaic current.
2. Neurons without axons or dendrites were studied in patch-clamp experiments.
3. To investigate the role of anodic and cathodic currents in the stimulation, rising and falling ramps were used in patch-clamp experiments.
4. Extracellular stimulation was examined and compared for both current-clamp and voltage-clamp experiments.

Therefore, the effects of the external electric field on the cell membrane could be studied in two similar equivalent models for cell-attached and cell-free membrane parts as shown in Fig. 2.15. The only difference is the cell attached resistance is high, as the ion channels are closed in the attached membrane. The attached part is connected through capacitance junction to the voltage source, while the free membrane is connected from outside to the solution medium and from inside to the voltage source through the cell-attached equivalent model and junction capacitor.

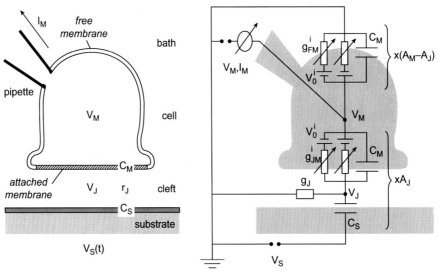

FIGURE 2.15

Electrical circuit model for capacitive stimulation of a nerve cell grown on top of a metal/oxide capacitor structure [13].

In Fig. 2.15, when an alternative voltage V_s is applied to the metal/insulator substrate with an area capacitance of C_s, a current flows through R_J junction resistance and an area-specific capacitance C_M. This leads to an extracellular potential drop over junction V_J with respect to bath solution (which is already grounded) and a change in intracellular potential as well. Rising voltage ramp is used to depolarize free membrane and triggering action potential. A falling ramp is applied to hyperpolarize the attached membrane, such that it can depolarize the free membrane and trigger an action potential through activation of ion currents.

Polarization of attached part of cell membrane is investigated under the current clamp by applying an extracellular voltage ramp signal. Fig. 2.16A shows current-clamp measurements for extracellular rising and falling voltage ramp stimulations. As the input is a ramp voltage, the intracellular potential will be a pulse square (see Fig. 2.16B), which is the time derivative of the ramp signal (as capacitance and resistance of the substrate acts like a derivator), and also the membrane current

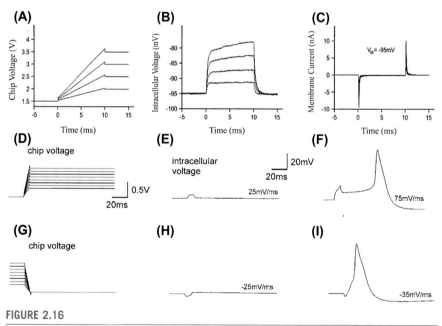

FIGURE 2.16

Voltage-clamp and current-clamp measurements for studying photocapacitive stimulation efficacy. (A) Input voltage applied to the capacitor. (B) Intracellular membrane potential. (C) Membrane current. (D) Input voltage applied to the capacitor. (E) Recorded intracellular voltage in current clamp. (F) Elicited action potential upon depolarization of the membrane through capacitive stimulation of the cell membrane. (G) Input falling applied voltage to capacitor. (H) Hyperpolarized cell upon applied falling input voltage. (I) Elicited action potential upon hyperpolarization of the attached membrane with falling input voltage [13].

is the time derivative of the membrane potential, so an impulse response (see Fig. 2.16C). Applying a raising voltage (see Fig. 2.16C,D and F) leads to the depolarization of the free membrane where ion channels are activated, and it results in action potential generation. Upon applying a falling voltage (see Fig. 2.16G,H and L), the potential will increase at the attached membrane and causes depolarization, which consequently activates Na^+/Ca^+ ion channels and firing of action potential. In addition, the falling voltage hyperpolarizes the free membrane as it acts passive. The dynamic of the initial phase of the stimulation is described as follows:

$$A_M c_M \frac{dV_M}{dt} + A_J \sum_i g_{JM} \left(V_M - V_0^i\right) \approx -A_J i_{JM}^{cap}(t) \qquad (2.12)$$

where V_M, A_J, c_M, V_0, g_{JM}, and JM are the membrane potential, junction area, membrane capacitance, resting potential, attached membrane conductance, and injected capacitive current.

Upon applying a rising voltage with the slope of $\Delta V_s / \Delta t_s$, the attached membrane is hyperpolarized such that sodium and calcium channels are inactivated. In this case, the attached membrane is hyperpolarized, and it acts like a passive coupling component, while the free membrane is depolarized. In the following equation, membrane potential is obtained from the resulting capacitive injected current:

$$(A_M - A_J) \left[c_M \frac{dV_M}{dt} + \sum_i g_{FM}^i \left(V_M - V_0^i\right) \right] \approx -A_J i_{JM}^{cap}(t) \qquad (2.13)$$

where $A_M - A_J$ is free membrane area and $(A_M - A_J) g_{FM}^i$ is free membrane conductance.

2.13 Anode break excitation

In some classes of cells, having another type of cation channels enables them to fire action potential with hyperpolarization [15–17]. This ion channel is activated slowly by hyperpolarization, leading to a growing inward current which is called I_h, I_f, I_Q, and I_{Ar}. For instance, in cardiac pacemakers and Purkinje fibers, hyperpolarization leads to excitation. I_h channels open at negative potentials similar to inwardly rectifying potassium (k_{ir}) channels, and they can be blocked by Cs^+ and Rb^+ ions [18]. These ion channels are permeable to both Na^+ and K^+ cations; therefore, they have linear I-V characteristics, with a negative reversal potential at approximately $-20mV$. Their channels have a steeply voltage gate-dependent activation and inactivation with sigmoid time course. As they pass an inward current after opening, they cause slow depolarization if the membrane potential becomes very negative. I_h plays a role in initiation of a recovery after a strong inhibition in central neurons. There are other examples of application for these channels, such as in photoreceptors of the eye, where I_h inhibits the effects of bright light. And in hearing, I_h or pacemaker current has been shown to contribute slowing of pacemaker depolarization. So, in contrast with the normal conditions under which we stimulate a neuron with depolarization, hyperpolarization of the cell causes firing of action potentials. In this case, after removal of hyperpolarization in a cell, decrease of potassium conductance and current has been explained to remove activation particle (n), and it results in increased inward sodium current leading to cell depolarization at membrane potential $V = 0$.

2.14 Photocapacitive stimulation of cells by photoswitch (Ziapin2)

Mattia Lorenzo et al. developed a light-sensitive azobenzene compound (Ziapin2) to stably place into the plasma membrane and (based on transdimerization in the dark) modulate cell capacitance and stimulate the cells through thinning effect [19]. The membrane capacitance (C_m) is calculated from the membrane characteristic features such as area (A) and thickness (d) through the following equation:

$$C_m = k_d \varepsilon_0 \frac{A}{l} \tag{2.14}$$

From the first derivative, we can find the contribution of the different parameters in the membrane capacitance changes:

$$dC_m = dk_d \varepsilon_0 \frac{A}{l} - k_d \varepsilon_0 \frac{A}{l^2} dl + k_d \varepsilon_0 \frac{dA}{l} \tag{2.15}$$

where C_m, K_d, ε_0, A, and l are cell capacitance, dielectric constant, permittivity of free space, surface area, and plasma membrane thickness, respectively. Considering that the first term in Eq. (2.15), change in area (dA) is indistinguishable, two other factors that can induce capacitance changes by the insertion of the photoswitch molecule are variations in dielectric constant, dk_d, and/or a change in thickness, dl. Therefore, the increase in the thickness (dl) in the second term of Eq. (2.15) is expected to decrease the capacitance.

Fig. 2.17A shows the patch-clamp measurements to study the effect of Ziapin2 in primary hippocampal neurons. In the dark condition, presence of the photoswitch in the cell membrane has been shown to increase cell membrane capacitance significantly (from 32.4 ± 1.7 pF to 53.3 ± 6.0 pF; see Fig. 2.17.B), while it does not have significant effects on other passive membrane properties. Under light condition, Ziapin2-loaded neurons were shown to have a fast and significant capacitance decrease (mean decrease $\pm$ s.e.m.: 6.5 ± 1.1 pF) and slowly return to the preillumination level (Fig. 2.17.B).

Current-clamp measurement for Ziapin2-labeled neurons has shown a biphasic change of the membrane potential upon photostimulation with an early hyperpolarization, with a coincidence with the capacitance change, and last by a delayed

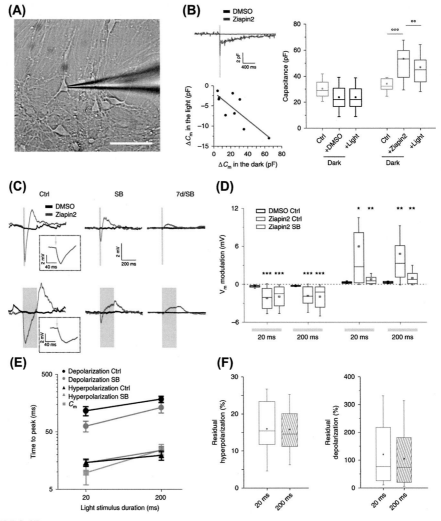

FIGURE 2.17

(A) In vitro whole-cell patch-clamp experiments for incubated hippocampal neurons with Ziapin2 by whole-cell patch clamp. Scale bar: 20 μm. (B) Left: Averaged capacitance measurements of neuron pulse labeled with either DMSO (0.25% v/v; black traces) or Ziapin2 (5 μM in DMSO; red traces), washed. Current-clamp configuration has been used in the presence of SB before and after light stimulation (470 nm; 18 mW/mm^2; cyan-shaded areas). The bottom panel shows single-cell correlation between the capacitance increase in the dark upon Ziapin2 addition (x axis) and the phasic capacitance decrease as a result of photostimulation (y axis). (C) Whole-cell current-clamp measurements from neurons incubated with either 0.25% (v/v) DMSO (black traces) or 5 μM Ziapin2 in DMSO (red traces) in the absence (Ctrl) or presence of SB and after 7 d of incubation in the presence of SB (7d/SB). The duration of the photostimulation (20 and 200 ms) is shown as a cyan-shaded area (470 nm: 18 mW/mm^2). (D) Box plots of the peak illustrates the alternation of hyperpolarization (left) and peak depolarization (right) in primary neurons exposed to DMSO/Ziapin2 and subjected to 20/200 ms photostimulation in the absence (Ctrl) or presence of SB. (E) The effect of light durations and membrane capacitance changes on the time-to-peak hyperpolarization, depolarization for different light stimulus durations under Ctrl, and synaptic block conditions. (F) Persistence of the photostimulation over time. Long-term photostimulation response of loaded Ziapin2 led to hyperpolarization (left) and depolarization (right) observed 7 d in the presence of SB, which are shown in percentage of the corresponding effects measured acutely after Ziapin2 loading. Box plots are shown for both 20 and 200 ms light stimuli [19].

depolarization with similar amplitude (Fig. 2.17C and D). While the hyperpolarization peak occurred with similar latency with 20 or 200 ms (for two different light durations of 20−200 ms), the peak depolarization shows a delay with 200 ms stimuli (Fig. 2.17E). In addition, blocking excitatory and inhibitory synaptic transmission enables us to prevent mixing the intrinsic effects and the effects of the reverberant network of synaptic connections. Fig. 2.17C and D shows that synaptic blockers (SB) do not affect the magnitude and timing of hyperpolarization ($P = .79$ and $P = .69$, Mann-Whitney U test). However, they cause the amplitude of the depolarization to decrease significantly (Fig. 2.17C and D; $P < .01$ for both 20 and 200 ms stimuli, Mann-Whitney U test). These results demonstrate that the peak hyperpolarization response generated by the presence of Ziapin2 is a result of an intrinsic response of the neuron. However, enhancement of depolarization, already present in synaptically isolated neurons, is enhanced by active synaptic transmission.

2.15 **Photovoltaic stimulation**

Photoelectrical stimulation of neurons is a promising strategy, due to its remote accessibility, and it has been considered a good alternative to optogenetics. In addition, it has attracted much attention in neuroscience and neuroprothesis, both in neuronal circuit investigation and curing of disorders such as retinal degeneration and death of retinal photoreceptors [1,20−23]. In this respect, there have been many efforts to employ different photovoltaic materials as artificial photoreceptors for visual prosthesis since the 1950s, when a photosensitive selenium cell was installed behind a patient's retina and resulted in phosphine detection after light illumination, until very recently, where conjugated organic polymers were applied for restoring vision in a rat model of degenerative blindness [24−28]. Fig. 2.18 shows the photoelectrical stimulation of cells based on QD dye synthesized and organic polymer solar cells.

Here, we review different photoactive materials and their working principles for neural stimulation.

2.15.1 **Quantum dots**

Among different photovoltaic materials, semiconductor nanocrystals (SNCs) have proven to be a good candidate because they may have high quantum photoelectrical efficiency and significant tunable optoelectronic properties, and their surface can be modified easily with different ligands and chemicals based on different favorite bindings to biological particles. In addition, the usage of SNCs in nanomedicine for in vivo molecular and cellular imaging, in targeting neuronal cells with QD-based multiphoton fluorescent pipettes, drug delivery, and in vivo cancer targeting has been reported before [31−35] a few times for photostimulation of neurons. However, the main issues of applying QDs practically as a neuron stimulator were

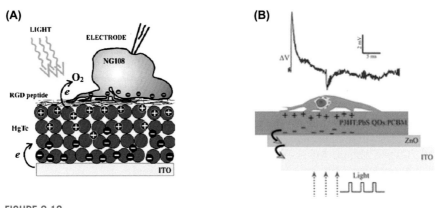

FIGURE 2.18

Quantam dot (QD)-based photoactive structures for photoelectrical stimulation of neurons. (A) Schematic of a QD-based photoelectrode structure [29]. (B) Schematic of an organic photoactive structure for photocapacitive stimulation of cells [30].

reported to be the high input light power needed for elucidation of action potential in previous studies and the lack of a practical model for neuronal modulation for clinical treatment based on the SNCs.

The first attempt at using QDs to communicate with neurons was in 2001 by Jessica Winter et al., who explained connection of QDs to neurons via antibody- and peptide-biding techniques [36]. Four years later, they showed photostimulation of rat neuron cells through Cds QDs. In 2007, Todd C. Pappas et al. described photocurrent mechanism for HgTE QDs as a tool for photostimulation of neuron cells [37]. And finally, Katherine Lugo et al. reported remote excitation of cancer and neuronal cells via CdTe and CdSe QDs based on induced photodipoles in QDs [38]. Although the studies showed the capability of QD interfaces for photostimulation of neurons, the light level used for stimulation has been demonstrated as much higher than practical level, about 800 mW/cm^2. Moreover, reviewing the early history of using QDs in energy harvesting devices such as solar cells and water-splitting devices, QDs have shown the potential to enhance the highest attainable thermodynamic conversion efficiency of photon conversions in solar cells up to 66% by using hot photogenerated charges, producing a higher level of photovoltage and photocurrent [39]. For example, in such solar cells, there have been several types of SCNCs like InP [40], Cds, CdSe, and PbS successfully studied already, some of which were used for neuronal stimulation.

Herein, possible mechanism of cell stimulation with QDs has been introduced. The QD-based photoelectrodes are made based on standard methods such as self-assembly for water splitting or other photovoltaic applications [40,41]. Fig. 2.19A shows a QD-based photoelectrode under measurement with a patch-clamp setup. As shown in the figure, the ITO conductive glass is connected to the ground of the amplifier and Ag/AgCl electrode immersed in the capillary glass pipette is

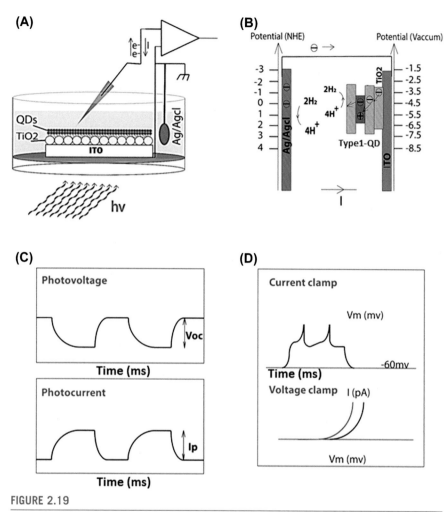

FIGURE 2.19

(A) Schematic of the photoelectrical stimulation setup with patch-clamp measurements. (B) Charge transfer mechanism in an Type 1 QD based photocathode; (C) measurements of photovoltage and photocurrent from the photocathode under light illumination with current and voltage-clamp technique; and (D) membrane potential and current measurement of a single cell grown on top of the photoelectrode.

collecting electrons from the surface of the polymer. In this manner, a closed-loop circuit enables measuring of the photocurrent from the photoactive layer. The photoelectrode structure consists of a titanium oxide (TiO_2) nanoparticle layer on indium thin oxide (ITO) conductive glass substrates, which facilitates electron transfer to the ground electrode. QDs are connected through ligands such as mercaptopropionic acid to the TiO_2 layer. After light absorption, they can separate electrical charges [2]. Fig. 2.19B illustrates ionic charge transfer for an TiO_2/Type 1 QD photoelectrode. This diagram describes energy diagram of the materials based on a hydrogen standard electrode system. Upon light illumination, the absorbed light in the QD layer will generate photoinduced charges. Then photoinduced electrons tend to move to the aCSF solution medium as the energy levels of the materials suggest (the charge transfer of the charges to the solution is dependent on Fermi energy level of the photoelectrode and the reduction/oxidation energy of the molecules in Faradaic reaction) [42], while holes move to the ITO substrate [42]. Since the electron transfer from the Type 1 QD core shell to the TiO_2 layer can be mediated by trap states in the TiO_2 layer, electron charge transfer could be easier from the TiO_2 layer to the QD layer and to the ionic water aCSF solution.

Photoelectrical characteristics of the photoelectrodes can be investigated by measuring the photocurrent in the aCSF medium. Fig. 2.19C top illustrates a typical photovoltage recorded by current-clamp technique. As the ITO/TiO2/Type 1 QD is a photocathode, the open voltage circuit is expected to be positive (in respect to the Ag/AgCl reference electrode), while the photocurrent is negative (Fig. 2.19C bottom). Fig. 2.19D shows expected membrane potential (up) and membrane conductance of a cell grown on top of a QD-based photoelectrode. Membrane conductance is also expected to be increased by photoinduced voltage and current as shown in Fig. 2.19D bottom. The red line is the measured conductance upon light illumination, which shows resting membrane potential has been changed as it has already opened some of the ion channels.

2.15.2 Organic semiconducting photocapacitors

Organic photovoltaics employ organic semiconducting polymers for converting light to electrical power [43]. Poly(3-hexylthiophene) (P3HT) is one of these conjugated organic polymers that is widely used to study the photophysics of the organic solar cell devices and in bioelectronic interfaces. This polymer can absorb light in the visible spectrum, whose absorption has a peak between 500 and 550 nm (green). This suggests that the visible wavelength of 520 nm (green light) for stimulation will lead to maximum absorption and free electrical charge generations in the polymer. Generally, upon light illumination, photoinduced charges and electron-hole couples (excitons) can be generated inside the polymer. While the P3HT is an p-type semiconductor (electron transport layer), the electrons in the polymer tend to enter the external solution medium, and the holes' direction will be toward the ITO side based on the energy band diagram in hydrogen standard energy system. Furthermore, in nature, without any closed-loop circuit, the photoinduced charges in the polymer generate surface potential and electric field decay from the medium interface because of the double layer effect. Also, these materials have unique optoelectronic properties and biocompatibility, which enables natural mimicking of the photoreceptors' performance. One example is organic interfaces for photostimulation of the retinal cells aimed to restore vision in case of photoreceptor degeneration based on P3HT:PCBM hybrid [1]. The organic polymeric particles also have been investigated for controlling cells' activity with light. They can be attached to the cell membrane surface with proper functionalization or injection inside the cell to alternate Ca^{2+} signaling via reactive oxygen (ROS) generation [44,45].

P3HT has been shown to be effective in neural stimulation, but the membrane potential displacement under light illumination was found to be just 1 mV maximum and Faradaic charge transfer is also involved based on the voltage-clamp recordings. In this regard, a recent study revealed that with engineering of the band energy diagram of the photoactive layer through adding an interlayer (ITO/ZnO/P3HT:PbS-QDs:PCBM [Phenyl-C61-butyric acid methyl ester]), a capacitive stimulation with high membrane potential displacement of 10 mV can be obtained [30]. To control the Faradaic and capacitive behavior of the interface, three different substrates of normal ITO/P3HT:PbS-QDs:PCBM, ITO/ZnO/P3HT:PbS-QDs:PCBM, and molybdenum oxide (**MoOx**) layer/P3HT:PbS-QDs:PCBM were investigated. Fig. 2.20 shows the biointerface structure and band energy

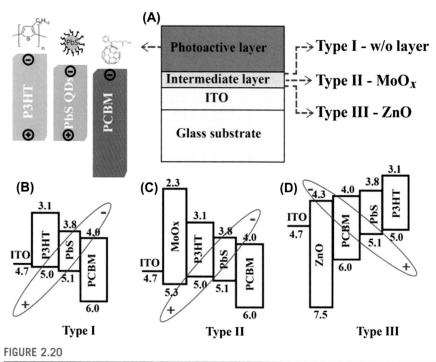

FIGURE 2.20

(A) The structure of photoelectrode device (Types I, II, and III). In all of them, P3HT, PbS QDs, and PCBM blend is the photoactive layer. (B) Energy band diagram for ITO/P3HT: PbS QDs:PCBM (Type I) with the photoinduced exciton after light illumination, (C) ITO/ MoOx/P3HT:PbS QDs:PCBM (Type II) with the photoinduced exciton after light illumination, and (D) ITO/ZnO/P3HT:PbS QDs:PCBM(Type III) with the photoinduced exciton after light illumination [30].

diagram for different substrates. If we consider this structure as a photocapacitor, the exciton (separated electrons in the conduction band energy of the first layer and holes in valence band energy of the last layer of the biointerface) in the biointerface layer has been shown to be as negative charges at ITO side and positive charges at last layer side (see Fig. 2.20B). Adding **MoOx** layer does not change the exciton direction, but because of the configuration of energy diagram, it leads to the lower amplitude of the exciton as a result of weaker charge transfer. Although, in Fig. 2.20C, adding **ZnO** layer leads to a change in the exciton direction. Fig. 2.21A shows the photocurrent setup, in which ITO substrate is connected to the ground Ag/AgCl electrode to create a closed-loop circuit with aCSF solution. Fig. 2.21B shows a dominant Faradaic current generated by ITO/P3HT:PbS-QDs: PCBM layer, while in Fig. 2.21C adding **MoOx** decreases this Faradaic current significantly, and finally, in Fig. 2.21D, with **ZnO** layer, the behavior of the biointerface becomes capacitive.

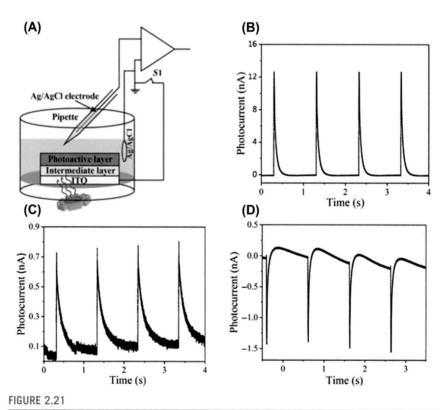

FIGURE 2.21

The photocurrent recordings for three different Faradaic and capacitive substrates. (A) Schematic of photocurrent patch-clamp setup. (B) Recorded photocurrent of ITO/ P3HT:PbS QDs:PCBM photoelectrode. (C) Recorded photocurrent of ITO/(MoOx) layer/ P3HT:PbS QDs:PCBM substrate. (D) Recorded photocurrent of ITO/ZnO/P3HT:PbS QDs: PCBM photoelectrode [30].

Fig. 2.22A shows the schematic of the photostimulation measurement with the patch-clamp amplifier. However, for the photocapacitive stimulation, an open loop is required, and so the ITO conductive layer is not grounded for the photostimulation measurements. Fig. 2.22B shows the membrane potential displacement upon 10 ms light illumination. Fig. 2.22C illustrates I-V characteristic of an immature SHSY-5Y cell recorded in voltage-clamp mode, implying that cell has a resting membrane potential about −30mv and a membrane conductance of approximately 66MΩ. In addition, to test the performance of the interface for different light intensities, membrane depolarization levels were measured with five different input light powers as shown in Fig. 2.22D.

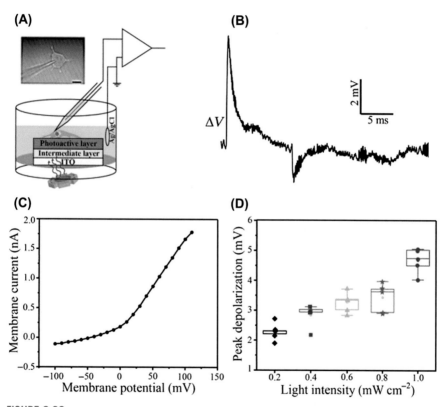

FIGURE 2.22

(A) Analysis of SH-SY5Y cells' responses to 10-ms light pulses on Type III photoelectrode architecture (ITO/ZnO/P3HT:PbS QDs:PCBM). (A) Schematic of the whole-cell patch-clamp recoding configuration of the photoelectrode in freestanding mode. The cells are grown on the photoelectrode and adhere to the polymer surface. A LED at far field is used to illuminate and activate the photoelectrode. (B) Current-voltage characteristics of a typical SH-SY5Y cell under dark. (C) Membrane potential variation upon light illumination (10 ms, 1 mW/cm^2). (D) Peak depolarization recorded for different light intensities (N = 5) [30].

2.15.3 Perovskite

Due to high solar to electric power conversion efficiency (PCE) in hybrid organic/inorganic halide perovskite photovoltaics and their low-cost fabrication, there have been many studies over the past decade on this material. Increase of PCE from 3% to 22% of perovskite solar cells during few years is unprecedented in the field of photovoltaics [46]. In spite of exciting optoelectronic properties of perovskite, two major challenges of its stability in liquid and its toxicity have prevented application of this material in photostimulation. Recently, the aforementioned issue has been solved by encapsulation of perovskite by a P3HT:polydimethylsiloxane layer to protect this material in cell medium, and its biocompatibility was enhanced. The first-time use of perovskite ($CH_3NH_3PbI_3$) as an ultrasensitive near IR interface, which enables photoelectrical stimulation of cells at a very low light intensity levels, was developed by Aria et al. in 2019 [47]. This biointerface can work up to near-IR, which facilitates using longer wavelengths for photostimulation of cells (Fig. 2.23).

In addition, tuning of the energy band gap of perovskite (MAPbX3) over the whole visible spectral range by adjusting the halide composition is one of the key advantages of this material. Based on this important advantage, color selection with energy band gap tuning enabled full-color nondissipative imaging based on

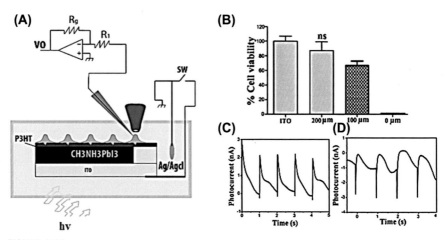

FIGURE 2.23

Schematic of photostimulation setup for a perovskite interface. (A) Patch-clamp configuration to measure photovoltaic properties of the perovskite substrate and also photostimulation experiments; perovskite is totally dark as it absorb the light up to near-IR spectrum, (B) cell viability of the SHSY-5Y cells grown on top of the encapsulated interface for different thickness of the encapsulated layer, (C) Faradaic photocurrent response of the noncontinuous perovskite layer, and (D) capacitive photocurrent generated from the continuous perovskite layer [47].

grown perovskite crystals, which was recently reported to closely resemble the cone cells in the human retina [48]. Moreover, operating in both narrowband and broadband regimes has made this material very promising for use as artificial photoreceptors for retinal prosthesis [49,50].

Fig. 2.23 shows photostimulation experiment and the process of patch clamp for a totally dark substrate of perovskite (Fig. 2.23). As the material absorbs all the light, observing grown cells on top of the biointerface is not easy, and so as the figure shows that a small part of the substrate has been left uncoated to provide enough visibility at the border of coated/uncoated part. In addition, cell viability assessment to test biocompatibility of the interface has been performed to make sure that cells are healthy (see Fig. 2.23B). In Fig. 2.23C and D, two Faradaic and capacitive photocurrent responses were obtained from two noncontinuous and continuous perovskite layers. In principle, based on the energy band diagram of the biointerface structure, we expect a capacitive current, while because of the voids at perovskite thin film in microcrystals, Faradaic leakage is dominant. To decrease the effect of the Faradaic contribution, a more continuous perovskite thin film was prepared by using multilayer spin coating of precursor solution for the same thickness, and finally a photocapacitive current was obtained as shown in Fig. 2.23D.

2.16 Photothermal stimulation

Photothermal stimulation of cells has been used for both inhibition and activation of the neurons. As the membrane capacitance relates to spatial distribution of charges around the membrane, any perturbation of the charge balance can lead to the change of membrane capacitance. While the optical local heat generation leads to redistribution of the electrical charges or charge density at the surface of the cell membrane, which modulates membrane capacitance, with long-term exposure light illumination, the generated local heat via metal or polymeric intermediators can increase membrane conductance and hyperpolarize the cells. Moreover, another approach for thermal stimulation is through activation of warmth-sensitive transient receptor potential (TRP) channels, as involved in body temperature regulation. In recent studies, it has been shown that TRPV1 ion channel in HEK 293T cells are responsible for depolarization of cells under photothermal stimulation mediated with P3HT [44].

2.16.1 Thermocapacitive stimulation

Using infrared (IR) light with wavelength >1.5um has enabled therapeutic potential as it remotely stimulates neurons and muscles without any genetic/chemical modifications or intermediators, although different mechanisms have been proposed in the literature for IR stimulations. Rapid temperature increase in tissue [51], which has been assumed to excite ion channels, activate intracellular messengers, form membrane pores, and excite TRPV channels and cause increase in ionic conductances, has been reported before [51−53]. Recently, one study revealed that thermocapacitive stimulation of neurons can be induced by a near IR or IR laser beam without any intermediator. In this study, Mikhail et al. showed IR light could excite cells through an electrostatic mechanism by altering cell capacitance, and it has been shown to be invasive and reversible [54]. Fig. 2.24A shows an equivalent electrical model of a cell membrane in the passive mode, in which its capacitance is temperature dependent. Voltage-clamp technique can be used to measure the conductance (membrane current) of an artificial lipid bilayer under illumination with an infrared laser pulse. Membrane capacitance is measured based on the impedance analysis of the cell membrane. Fig. 2.24B is the time response of the induced current under laser pulse illumination in voltage clamp (-200mv to 200 mV), and Fig. 2.24C is the measured laser-induced charges for H_2O and D_2O mediums. As shown in the figure, the induced charges in water medium is much higher than in D_2O, confirming the fact that water acts as a primary chromophore (it has an absorption coefficient of 60.6 cm^{-1} at 1889 nm) and converts the

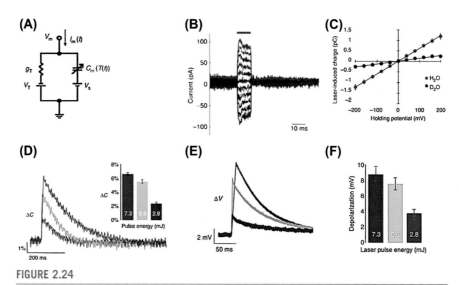

FIGURE 2.24

Depolarization of artificial lipid bilayers based on induced electrical membrane capacitance changes by infrared light stimulation. (A) A passive model of an electrical equivalent circuit for cell membrane with a temperature-dependent capacitance (B) recorded the I-V in voltage-clamped artificial lipid bilayer in symmetric NaCl solution; (C) alternation of membrane capacitance for different laser pulse durations of 0.2, 0.5, 0.75, and 1 ms; (D) membrane potential change for different laser pulses; (E) induced depolarization for three different laser powers; (F) firing of action potentials of oocytes coexpressing sodium and potassium channels in voltage-clamped mode. Red signal is with the black line without laser illumination, and the bars indicate the time of the laser illuminations [54].

IR light to energy for cell excitation. In addition, the response of the photoinduced capacitance has been shown in Fig. 2.24D for three different light intensities resembling the absolute temperature around the lipid bilayer. The resulting voltage response is measured in current-clamp mode, and Fig. 2.24E and F indicates higher power laser illumination induces more membrane capacitance, which causes higher time constant (slower rise time) and higher membrane depolarization (more charge accumulation) at the longer times.

2.16.2 Organic polymers

Photothermal effect in semiconductors originates from the diffusion of photogenerated charge carriers (electrons) in the material and its conversion to molecular lattice vibration (phonons) or heat dissipation. Organic semiconductor polymers such as P3HT have been employed for photothermal stimulation based on increase of temperature at its surface, and so it alternates the electrical cell membrane properties such as electrical capacitance and resistance. Fig. 2.25 shows that the temperature

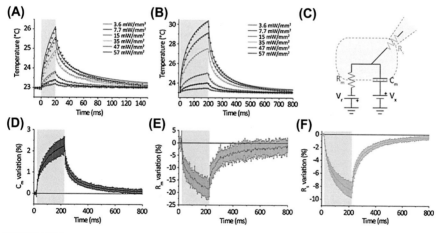

FIGURE 2.25

Increase in temperature measured (open circuit) in the bath in close proximity to the polymer (P3HT on glass) surface for 20 ms (A) and 200 ms (B) pulses at different light intensities. Solid lines represent numerical simulation of the thermal diffusion problem. (C) Equivalent circuit representation of a cell membrane in a patch-clamp measurement. R_m and C_m are the membrane resistance (that includes the effect of all HEK-293 ion channels) and capacitance, respectively, R_s is the series resistance of the patch, V_r is the reversal potential, and V_x is a term needed to take into account the asymmetries between inner and outer membrane surface charges and ion distributions. (D,e,f) Time evolution of membrane capacitance (D), membrane resistance (E), and series resistance (F) during illumination (200 ms, 57 mW/mm^2—cyan rectangle), measured for $n = 39$ cells (error bars represent standard deviations) [55].

arises as a result of light illumination for different light power intensities and exposure times. As Fig. 2.25 illustrates, cell membrane conductance and capacitance are varied by light illumination. Based on the results, induced temperature increases cell membrane capacitance by 2% and reduces membrane resistance by 20% consequently in HEK cells grown on top of P3HT polymer. It is obvious that increase of cell capacitance with the same resting membrane potential results in increase of intracellular potential to balance the charge equilibrium for the intracellular and extracellular environment and sets the membrane potential to rest again. On the other hand, decrease in resistance and hyperpolarization could be because of activation of the potassium ion channels, as the resting potential is close and set by the potassium voltage-gated ion channels.

The membrane capacitance is proportional to the temperature as follows [55]:

$$C_m(T) = C_m(T_0).[1 + \alpha_C(T - T_0)] \tag{2.16}$$

In which α_C is the relative increase for 1 °C temperature change and T_0 the baseline room temperature. To calculate R_m, a temperature-dependent expression with temperature coefficient of Q_{10} is used as follows [55]:

$$R_m(T) = R_m(T_0).Q_{10}^{-\frac{T-T_0}{10}} \tag{2.17}$$

Temperature not only affects the membrane resistance but also a new equilibrium condition, as different channels have different temperature coefficients [55].

$$V_r = \frac{RT}{F} \ln\left(\frac{P_{Na^+}[Na^+]_{out} + P_{K^+}[K^+]_{out} + P_{Cl^-}[Cl^-]_{in}}{P_{Na^+}[Na^+]_{in} + P_{K^+}[K^+]_{in} + P_{Cl^-}[Cl^-]_{out}}\right) \tag{2.18}$$

The approximate relationship for the reversal potential versus temperature can be obtained by Ref. [55]:

$$V_r(T) = V_r(T_0).\left(\frac{T}{T_0}\right)^{\alpha V} \tag{2.19}$$

To consider the effect of photothermal and photoelectrical stimulation by P3HT separately, one can use the following equation including two terms of the induced photovoltaic and photothermal membrane capacitance [56]:

$$i_C = \frac{dQ}{dt} = C_m \frac{d(V - V_s)}{dt} + (V - V_s)\frac{dC_m}{dt} \tag{2.20}$$

where the V_s is the surface membrane potential. In fact, C_m depends on both surface membrane potential and temperature effects, so the second term can be expanded as follows [56]:

$$\frac{dC_m}{dt} = Cm\left(\frac{dC_m}{dT}.\frac{dT}{dt} + \frac{dC_m}{dV_s}.\frac{dV_s}{dt}\right) \tag{2.21}$$

2.17 Conclusion

In this chapter, general patch-clamp technique was introduced in a practical manner, and different types of Faradaic and capacitive neural stimulation were described. Then, using patch clamp for studying the mechanism behind the extracellular stimulation with a metal/insulator neural interface was explained. Therefore, it provides the guidance for extracellular stimulation with the patch-clamp setup and also for studying a typical capacitive neural interface. Moreover, two important types of photoelectrical and photothermal neural stimulations and the patch-clamp measurements to study stimulation mechanisms were explained in this chapter.

References

[1] Ghezzi D, et al. A polymer optoelectronic interface restores light sensitivity in blind rat retinas. Nature Photonics 2013;7(5):400−6.

[2] Bahmani Jalali H, et al. Effective neural photostimulation using indium-Based type-II quantum dots. ACS Nano 2018;12(8):8104−14.

[3] Carvalho-de-Souza JL, et al. Optocapacitance allows for photostimulation of neurons without requiring genetic modification. In: Use of nanoparticles in neuroscience. Springer; 2018. p. 1−13.

[4] Grahame DC. The electrical double layer and the theory of electrocapillarity. Chemical Reviews 1947;41(3):441−501.

[5] Adams D, Nonner W. Voltage-dependent potassium channels: gating, ion permeation and block. In: Potassium channels: structure, classification, function and therapeutic potential; 1990. p. 40−69.

[6] Moczydlowski E, Lucchesi K, Ravindran A. An emerging pharmacology of peptide toxins targeted against potassium channels. Journal of Membrane Biology 1988; 105(2):95−111.

[7] Johnston J, Forsythe ID, Kopp-Scheinpflug C. Symposium review: going native: voltage-gated potassium channels controlling neuronal excitability. The Journal of Physiology 2010;588(17):3187−200.

[8] Nichols C, Lopatin A. Inward rectifier potassium channels. Annual Review of Physiology 1997;59(1):171−91.

[9] Elliott A, Elliott J. Characterization of TTX-sensitive and TTX-resistant sodium currents in small cells from adult rat dorsal root ganglia. The Journal of Physiology 1993;463(1):39−56.

[10] Randall A, McNaughton N, Green P. Properties of voltage-gated Na+ channels in the human rhabdomyosarcoma cell-line SJ-RH30: conventional and automated patch clamp analysis. Pharmacological Research 2006;54(2):118−28.

[11] Ortmanns M. Charge balancing in functional electrical stimulators: a comparative study. In: 2007 IEEE international symposium on circuits and systems. IEEE; 2007.

[12] Merrill DR, Bikson M, Jefferys JG. Electrical stimulation of excitable tissue: design of efficacious and safe protocols. Journal of Neuroscience Methods 2005;141(2):171−98.

[13] Schoen I, Fromherz P. The mechanism of extracellular stimulation of nerve cells on an electrolyte-oxide-semiconductor capacitor. Biophysical Journal 2007;92(3):1096−111.

[14] Schoen I, Fromherz P. Extracellular stimulation of mammalian neurons through repetitive activation of Na+ channels by weak capacitive currents on a silicon chip. Journal of Neurophysiology 2008;100(1):346−57.

[15] Yanagihara K, Irisawa H. Inward current activated during hyperpolarization in the rabbit sinoatrial node cell. Pflügers Archives 1980;385(1):11−9.

[16] Spain W, Schwindt P, Crill W. Anomalous rectification in neurons from cat sensorimotor cortex in vitro. Journal of Neurophysiology 1987;57(5):1555−76.

[17] Bader C, Bertrand D. Effect of changes in intra-and extracellular sodium on the inward (anomalous) rectification in salamander photoreceptors. The Journal of Physiology 1984;347(1):611−31.

[18] Hille B. Ionic channels in excitable membranes. Current problems and biophysical approaches. Biophysical Journal 1978;22(2):283−94.

[19] DiFrancesco ML, et al. Neuronal firing modulation by a membrane-targeted photoswitch. Nature Nanotechnology 2020:1−11.

[20] Bareket-Keren L, Hanein Y. Novel interfaces for light directed neuronal stimulation: advances and challenges. International Journal of Nanomedicine 2014;9(1):65−83.

[21] Ghezzi D, Antognazza MR, Dal Maschio M, Lanzarini E, Benfenati F, Lanzani G. A hybrid bioorganic interface for neuronal photoactivation. Nature Communications 2011;2(1):1−7.

[22] Abdullaeva OS, et al. Photoelectrical stimulation of neuronal cells by an organic semiconductor−electrolyte interface. Langmuir 2016;32(33):8533−42.

[23] Martino N, et al. Optical control of living cells electrical activity by conjugated polymers. Journal of Visualized Experiments 2016;(107):e53494.

[24] Lo W. Implantable, fully integrated and high performance semiconductor device for retinal prostheses. Google Patents. 2005.

[25] Keller N, et al. Artificial retina that includes a photovoltaic material layer including a titanium dioxide semiconductor. Google Patents. 2015.

[26] Hu L, et al. Organic optoelectronic interfaces with anomalous transient photocurrent. Journal of Materials Chemistry C 2015;3(20):5122−35.

[27] Maya-Vetencourt JF, Ghezzi D, Antognazza MR, Colombo E, Mete M, Feyen P, Desii A, Buschiazzo A, Di Paolo M, Di Marco S, Ticconi F. A fully organic retinal prosthesis restores vision in a rat model of degenerative blindness. Nature Materials 2017;16(6):681−9.

[28] Tassicker G. Preliminary report on a retinal stimulator. British Journal of Physiological Optics 1956;13(2):102−5.

[29] Pappas TC, et al. Nanoscale engineering of a cellular interface with semiconductor nanoparticle films for photoelectric stimulation of neurons. Nano Letters 2007;7(2):513−9.

[30] Srivastava SB, et al. Band alignment engineers faradaic and capacitive photostimulation of neurons without surface modification. Physical Review Applied 2019;11(4):044012.

[31] Bruchez M, et al. Semiconductor nanocrystals as fluorescent biological labels. Science 1998;281(5385):2013−6.

[32] Andrásfalvy BK, et al. Quantum dot-based multiphoton fluorescent pipettes for targeted neuronal electrophysiology. Nature Methods 2014;11(12):1237−41.

[33] Gao X, et al. In vivo cancer targeting and imaging with semiconductor quantum dots. Nature Biotechnology 2004;22(8):969−76.

[34] Gao X, et al. In vivo molecular and cellular imaging with quantum dots. Current Opinion in Biotechnology 2005;16(1):63–72.

[35] Dubertret B, et al. In vivo imaging of quantum dots encapsulated in phospholipid micelles. Science 2002;298(5599):1759–62.

[36] Winter JO, et al. Recognition molecule directed interfacing between semiconductor quantum dots and nerve cells. Advanced Materials 2001;13(22):1673–7.

[37] Winter JO, et al. Quantum dots for electrical stimulation of neural cells. In: Biomedical optics 2005. International Society for Optics and Photonics; 2005.

[38] Lugo K, et al. Remote switching of cellular activity and cell signaling using light in conjunction with quantum dots. Biomedical Optics Express 2012;3(3):447–54.

[39] Grätzel M. Dye-sensitized solar cells. Journal of Photochemistry and Photobiology C: Photochemistry Reviews 2003;4(2):145–53.

[40] Zaban A, et al. Photosensitization of nanoporous TiO_2 electrodes with InP quantum dots. Langmuir 1998;14(12):3153–6.

[41] Kongkanand A, et al. Quantum dot solar cells. Tuning photoresponse through size and shape control of CdSe– TiO2 architecture. Journal of the American Chemical Society 2008;130(12):4007–15.

[42] Grimes C, Varghese O, Ranjan S. Light, water, hydrogen: the solar generation of hydrogen by water photoelectrolysis. Springer Science & Business Media; 2007.

[43] Günes S, Neugebauer H, Sariciftci NS. Conjugated polymer-based organic solar cells. Chemical Reviews 2007;107(4):1324–38.

[44] Bossio C, et al. Photocatalytic activity of polymer nanoparticles modulates intracellular calcium dynamics and reactive oxygen species in HEK-293 cells. Frontiers in Bioengineering and Biotechnology 2018;6.

[45] Zangoli M, et al. Engineering thiophene-based nanoparticles to induce phototransduction in live cells under illumination. Nanoscale 2017;9(26):9202–9.

[46] Correa-Baena J-P, et al. The rapid evolution of highly efficient perovskite solar cells. Energy and Environmental Science 2017;10(3):710–27.

[47] Aria MM, et al. Perovskite-based optoelectronic biointerfaces for non-bias-assisted photostimulation of cells. Advanced Materials Interfaces 2019:1900758.

[48] Yakunin S, et al. Non-dissipative internal optical filtering with solution-grown perovskite single crystals for full-colour imaging. NPG Asia Materials 2017;9(9):e431.

[49] Saidaminov MI, et al. Perovskite photodetectors operating in both narrowband and broadband regimes. Advanced Materials 2016;28(37):8144–9.

[50] Fang Y, et al. Highly narrowband perovskite single-crystal photodetectors enabled by surface-charge recombination. Nature Photonics 2015;9(10):679–86.

[51] Wells J, et al. Biophysical mechanisms of transient optical stimulation of peripheral nerve. Biophysical Journal 2007;93(7):2567–80.

[52] Richter CP, et al. Neural stimulation with optical radiation. Laser and Photonics Reviews 2011;5(1):68–80.

[53] Katz EJ, et al. Excitation of primary afferent neurons by near-infrared light in vitro. Neuroreport 2010;21(9):662–6.

[54] Shapiro MG, et al. Infrared light excites cells by changing their electrical capacitance. Nature Communications 2012;3:736.

[55] Martino N, et al. Photothermal cellular stimulation in functional bio-polymer interfaces. Scientific Reports 2015;5:8911.

[56] Carvalho-de-Souza JL, et al. Optocapacitive generation of action potentials by microsecond laser pulses of nanojoule energy. Biophysical Journal 2018;114(2):283–8.

Electrophysiological characteristics of neuron-like cancer cells and their applications for studying neural interfaces

3.1 Introduction

Neuron-like cells such as PC12, NG108, SHSY-5Y, and Neuro-2a cells are useful in neural stimulation studies. They have proliferation rates and can be multiplied, eliminating the need to use primary or explant neurons requiring animal sacrifices. In addition, cell lines provide sufficient homogeneous cells and easily controlled conditions for cell stimulation experiments. Another advantage of these cells is that their electrical excitability, such as the threshold for action potential (AP) generation, can be controlled through nerve growth factor (NGF) treatments and facilitates investigation of the effect of stimulation on a mature and immature cell.

Exploring the effect of neural stimulation with neuron-like cells needs patch-clamp studies, as the understanding of electrophysiological behaviors of these cells without interaction with the interfaces is crucial. For that, cells are grown on a normal substrate and voltage-clamp experiments to investigate I-V characteristic and current-clamp to study the threshold for AP generation are performed. Afterward, the same experiments on the cells grown on top of the interface are repeated without any external stimuli, to find if the electrophysiological behaviors of the cells are altered after attaching to the surface of the neural interface. Finally, cell stimulation experiments are utilized to examine the performance of the neural interface. Based on these experiments, researchers will understand the mechanism behind cell stimulation from the changes in membrane potential and ionic conductances under stimulation.

In this chapter, basic information on PC12, NG108, SHSY-5Y, and Neuro-2a cells is discussed. Then, basic patch-clamp measurements for cell lines are presented, so that researchers can determine which cell line can satisfy their requirements for their experiments. In addition, two studies in which they neuron-like cell lines have been employed to study the efficacy of neural interfaces are reviewed. Moreover, the protocol for mathematical modeling of AP shape based on thermodynamic approach has been explained.

Electrophysiology Measurements for Studying Neural Interfaces. https://doi.org/10.1016/B978-0-12-817070-0.00003-8

3.2 PC12 cells and differentiation to neuronal-like cells

PC12 is a cell line derived from rat pheochromocytoma cells and a classical neuronal cell model, as it has neuronal features such as AP generation when it undergoes treatment with NGF. Normally, PC12 cells cannot generate AP, but they already have potassium and calcium voltage-gated channels. After 2 weeks of NGF treatment, the number of Na^+ channels will increase drastically and be high enough to produce AP. NGF has two effects of increasing the number of functional Na^+ channels and inducing tetrodotoxin (TTX)-resistant Na^+ channels [1]. Moreover, NGF treatment affects microtubule-dependent extension and maintenance of axons [2]. Fig. 3.1A and B shows the morphological change of cultured PC12 cells before and after NGF treatment.

There are some studies of the excitability of PC12 cells through TTX-resistant Na^+ channels and resulted electrophysiological behaviors after differentiation. Rudy et al. showed that without TTX, above 80% of the cells were measured to produce APs as the one shown in Fig. 3.2A [3]. Among 98 NGF-treated cells under patch-clamp recording in the presence of 100 nM TTX, only 34 could generate APs, while the threshold for generation of AP was higher, with a slower rate of increase than those that were measured in the absence of TTX. However, in the presence of 2PM TTX, none of the cells was shown to produce APs. Therefore, active electrical responses could be adjusted by the TTX-resistant channels. The homogeneity of excitability and the efficiency of differentiation in cells treated with 100 nM TTX could be due to factors such as differences in how many of the channels are unblocked at 100 nM TTX, cell population, and confluency or differences in leak ionic currents. In their experiments, Rudy et al. showed that Na^+ channels in differentiated PC12 cells are similar to those in other voltage-gated undifferentiated cells. Along with being voltage-dependent and having a threshold in number of ion channels after NGF treatment for producing APs, these channels contain three sites at which toxins can interact: a site for binding TTX, saxitoxin (STX), and the binding

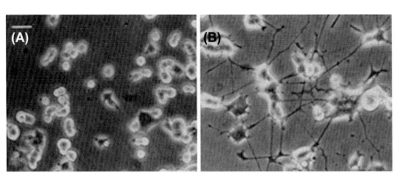

FIGURE 3.1

(A) Undifferentiated PC12 cells and (B) differentiated PC12 cells in 100 ng/mL NGF for 5 days [2].

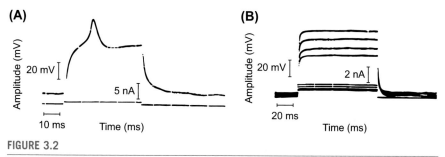

FIGURE 3.2

The current-clamp measurement to investigate excitability of differentiated PC12 cells without presence of TTX and with presence of TTX [1]. (A) AP recorded from a differentiated PC12 cell in current-clamp mode. (B) Inhibition of AP with presence of TTX: as it shows with injecting current in current-clamp mode, the differentiated PC12 cell shows depolarization but not the generation of AP.

site at which the blockage of ion flux can happen; a binding site for alkaloid toxins like veratridine and batrachotoxin leading to persistent activation of ion channel at voltages that closes or inactivates the channel; and a regulatory binding site for scorpion toxins, which works with alkaloid toxins in a cooperative way [4]. Their studies also on the interaction of Na^+ channels in PC12 cells with TTX and STX showed that Na^+ channels were qualitatively similar in PC12 cell membranes, while long-term exposure to NGF changes the number of this channels. An increase of 15- to 25-fold in channel numbers per cell or 5- to 7-fold per unit area of membrane was found. This was confirmed by studying ionic fluxes. In addition, it has been shown that inducing NGF increases STX binding and represents an increase in functional TTX-sensitive Na^+ channels, which are as the same as those in NGF-untreated PC12 cells. Furthermore, a subpopulation of channels with low affinity to TTX has been shown to exist. An increase in functional TTX-sensitive channels has been correlated to the magnitude found in binding studies. It was also explained that because the affinity of TTX-sensitive Na^+ channels for STX is approximately 10-fold lower than for TTX, STX binding in NGF-treated cells may include a small fraction of the population of TTX-resistant channels.

An accurate method to find the number of expressed ion channels after NGF treatment based on the patch-clamp technique is the measurement of ionic conductance. The sodium conductance directly relates total Na^+ conductance to a single Na^+ channel conductance, the total number of ion channels (including expressed ion channels after NGF treatment), and the fraction time when all channels are activated.

$$G_{Na} = gnf \tag{3.1}$$

where g is the single ion channel conductance, n is the number of Na^+ ion channels, and f is the time fraction.

3.2.1 Voltage-clamp and current-clamp measurement

To study PC12 cell stimulation on a biointerface, electrophysiological properties of PC12 cells are studied without an external stimulus signal through the interface. In this regard, voltage-clamp and current-clamp analysis are used to find membrane resting potential, ionic conductance, and the threshold for AP generations. Fig. 3.3A and B shows an optical microscopy image of PC12 grown cells and an approached PC12 cell that is under patch-clamp recording. Fig. 3.3.C shows that the I-V obtained from the voltage-clamp measurement shows a total membrane resistance equal to 125 MOhm and a resting membrane potential of about −35 mV, although these parameters may vary in time during the recordings. After measuring voltage-clamp, current-clamp recording is used to find the threshold for AP generation after applying different input injected currents. Fig. 3.3D shows the AP generated after injecting an input current of 360 pA.

3.2.2 Differentiation protocol for PC12 cells

For most of the reported experiments, NGF-treated PC12 cells are prepared as following: cells are grown in the presence of NGF [6] with a concentration of 50 mg/mL for minimum 2 weeks on collagen-coated dishes and then detached, and they are transferred onto polyornithine-coated 18 mm multiwell dishes [7] for an extra 24 h incubation in the presence of NGF. Cultures of non-NGF-treated cells could be similarly prepared but without NGF at any step.

3.3 NG108-15 cells

The NG108 is a hybrid of neuroblastoma/glioma cell line and has been used for studying neural interfaces [8,9], whole-cell biosensing [10], and toxin detection [11]. This cell line already has all voltage-gated ion channels of sodium, potassium, and calcium, and it can generate immature APs.

3.3.1 Voltage-clamp and current-clamp experiments

To use NG108 cells for studying neural stimulation, whole-cell recording of cells is measured (see Fig. 3.4). Time responses of the membrane current for different input applied voltages across the cell membrane are measured to obtain the I-V curve (see Fig. 3.4A and B). From I-V, resting membrane potential of about −45 mV and membrane resistance of 125 MOhm can be calculated. In addition, to find the time response of membrane potential and the threshold of injected current (depolarization) for AP generation, current-clamp technique is used. As Fig. 3.4C shows upon injecting a 10 pA a depolarization level of 25 mV, for 20 pA depolarization of 30 mV, and for 30 pA depolarization of about 30 mV but an immature AP with the amplitude of

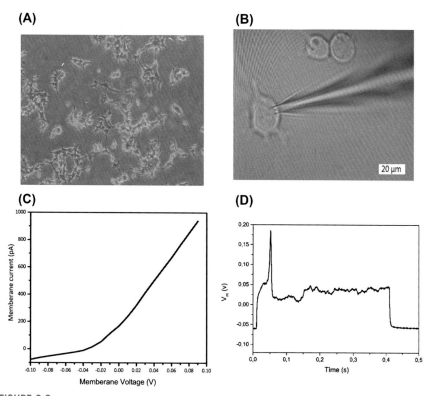

FIGURE 3.3

Optical image of NGF-treated PC12 cells and patch-clamp measurements. (A) Optical image of differentiated and undifferentiated PC12 cells, (B) a PC12 cell under patch-clamp recording, (C) I-V characteristic of a PC12 cell, and (D) action potential generated by injecting 360 pA current in current-clamp mode [5].

5 mV have been recorded. Moreover, a maximum amplitude for the immature AP has been shown to be achieved about 20 mV by injecting 200 pA (see Fig. 3.4D).

For producing mature APs, NG108 cells are treated with NGF for expression of voltage-gated Na^+ channels. Fig. 3.5 shows differentiation of NG108 cells after NGF treatment within 21 days. In bright field images, axonal development after 21 days in the cells and in fluorescent images with the choline acetyltransferase marker together confirms effective cell differentiation. Moreover, whole-cell results as shown in Fig. 3.6 for different treatment days explain the development of the AP shape for different differentiation levels. In Fig. 3.6, whole-cell current-clamp recordings for $n = 22$ cells show the resting membrane potential recorded at -34.9 ± 1.4 mV ($n = 22$). At Day 0, a small bump in the membrane potential after injecting a step current is observed. After 9 days its amplitude becomes stronger, and finally at Day 21, a burst of APs can be observed.

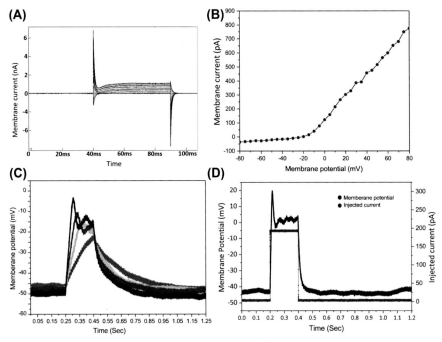

FIGURE 3.4

Voltage-clamp and current-clamp measurements of untreated NG108 cells. (A) Traces of voltage-clamp responses, (B) obtained I-V characteristic of from (A), (C) current-clamp measurement for membrane potential for different injected currents, and (D) maximum membrane overshoot (immature action potential) for an input injected current of 200 pA.

3.3.2 Sodium current and differentiation

Voltage-clamp recording of sodium ionic currents is employed with the current-clamp measurement of APs for the same cell to correlate the amount of increased Na^+ channels to differentiation level. For undifferentiated cells, Na^+ current density has been shown (see Fig. 3.7) to be -29.2 ± 3 pA/pF ($n = 10$), while after differentiation the level of sodium current reached -40.8 ± 3.3 pA/pF. After 21 days of differentiation, the amount of sodium current increased to -113.5 pA/pF for cells that generated APs. Complete inhibition of Na^+ current density by 1uM TTX in differentiated and undifferentiated cells suggested the existence of only TTX-sensitive Na^+ channels ($Na_v1.7$) in NG108 cells.

3.3.3 Protocol for NG108 differentiation

The cells are cultured in a serum-free medium made of DMEM, N2 supplements, 1 mM dibutyryl cyclic AMP, and antibiotics. After 0−21 days of differentiation, they can be used for experiments.

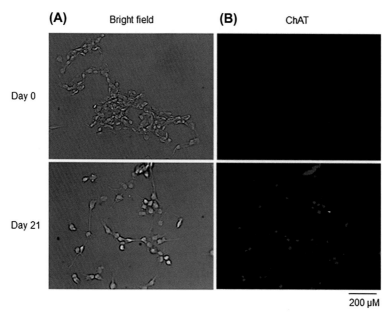

FIGURE 3.5

The optical and fluorescent microscope images for the undifferentiated and differentiated cells. (A) Optical microscope images from undifferentiated cells and (B) fluorescent image of the differentiated NG108 cells labeled by choline acetyltransferase (ChAT) marker [12].

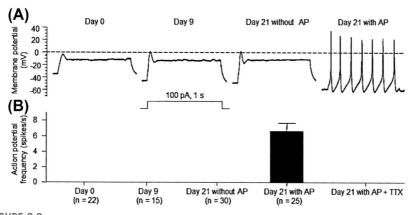

FIGURE 3.6

Current-clamp recording from NG108 cells. (A) The membrane potential response to a current injection (100 pA, 1 s) in current-clamp configuration. (B) Action potential (AP) generation in undifferentiated (Day 0) and differentiated NG108-15 cells. *TTX*, tetrodotoxin. Data are means ± SEM, *n* is the number of cells [12].

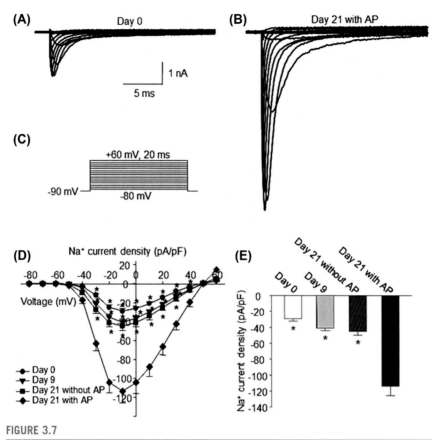

FIGURE 3.7

Voltage-clamp recording from undifferentiated and differentiated cells. (A) Sodium current of undifferentiated cells, (B) sodium current of differentiated cells, (C) sweep of voltage, (D) I-V characteristic of sodium channel for NG108 differentiated and undifferentiated cells, and (E) the level of Na^+ current density for different differentiation times [12].

3.3.4 Mathematical model for the shape of action potential analysis

Extracting mathematical models for excitable cells is advantageous because it helps us to understand and explain the effect of cell stimulation or inhibition from mathematical parameters. Different mathematical models based on the land mark work of Hodgkin and Huxley have been developed to relate the electrical properties of cells to the process of AP generation [10,13,14]. Hodgkin-Huxley is the widely used model in which time derivative of membrane potential is related to ionic currents and thus channel activation and inactivation of voltage-gated ion channels controlling ionic fluxes [15]. In this model as it was discussed in Chapter 1, the voltage and time dependence of the gates were described by empirical functions of the membrane potential. However, a more simplified functional form of the voltage dependence of the rate constants from thermodynamics has been introduced [16].

Simulation of ionic conductances and AP generation: Fig. 3.8 shows the flowchart for finding fitting parameters in the thermodynamic approach for modeling an AP based on the experimental results. It facilitates predicting the correlation of kinetic parameters of ion channels in forming AP and understanding the mechanisms behind cell stimulation and inhibitions. Based on the flowchart, voltage-clamp measurement is used to extract voltage-gated ion channel parameters, while current-clamp measurement is used to record APs and membrane capacitance. Next step is an iteration between the membrane current and potential equations to find the optimum fitting parameters that give a fine simulated AP close to the experimental one. Mohan et al. explained how, based on the APformalism, they could simulate AP shape of differentiated NG108-15 cells for toxin detection, and the following section reviews their study.

To obtain fitting parameters, the following steps are taken:

1. Using a computer program that fits parameters to the recorded data based on Eq. (3.2). Therefore, experimental results are obtained from recorded ionic currents versus time and membrane potential.
2. Extracting ion channel parameters from the voltage-clamp experiments: Ionic currents are recorded in voltage-clamp mode using the protocols explained in Chapter 2. Fig. 3.9A shows a patched NG108 cell in witch its sodium and potassium ion currents were recorded in voltage-clamp (as shown in Fig. 3.9B and C). The data are imported into a computer software like MATLAB program. Briefly, the total ionic membrane current has been described as follows:

$$I_{ionic} = I_{Na} + I_K + I_{Ca} + I_1 = \bar{g}_{Na}m^3h(V - V_{Na}) + \bar{g}_K n^4(V - V_K) + \bar{g}_{Ca}e^3(V - V_{Ca}) + \bar{g}_l(V - V_l) \tag{3.2}$$

where g_{Na}, g_K, and g_{Ca} and V_{Na}, V_K, and V_{Ca} are sodium, potassium, and calcium maximum conductances and potentials, respectively, and m, n, h, and e are the state variables.

   a. The dynamic state of variables is calculated as follows:

$$\frac{dm}{dt} = \frac{m_\infty - m}{\tau_m} \tag{3.3}$$

where m, n, h, and e are the steady-state values of the state variables and τ_m, τ_n, and τ_h are voltage-dependent time constants, respectively.

   b. According to the general thermodynamic formalism, the steady-state parameters and time constants can be calculated, as an example for the m state parameter, we have

$$m_\infty = \frac{1}{1 + \exp^{-(zF/RT)(V_m - V_{1/2})}} \tag{3.4}$$

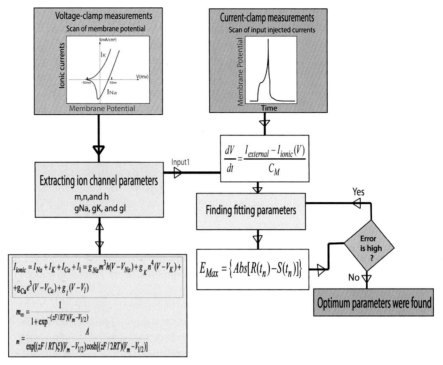

FIGURE 3.8

Flowchart for finding fitting parameters and simulation of action potential (AP) shape.
Voltage-clamp measurement is used to extract ion channel parameters, while current-
clamp measurements are employed to record APs. Extracted parameters from voltage-
clamp recordings are used in membrane potential equation and an iteration is used to find
fitting parameters. An error function is used to compare the experimental and simulated
AP for the optimum values of the fitting parameters and minimum achievable and
acceptable error.

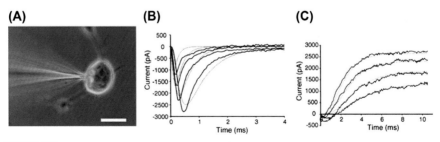

FIGURE 3.9

Patch-clamp of a single NG108 cell. (A) Optical microscope image of a patched cell,
(B) sodium ion currents recorded in voltage-clamp experiment, and (C) potassium
current recorded at four different membrane potentials of 0, 10, 20, and 30 mV [10].

and

$$\tau_m = \frac{A}{\exp[(zF/RT)\xi]\left(V_m - V_{1/2}\right)\cosh\left[(zF/2RT)\left(V_m - V_{1/2}\right)\right]} \quad (3.5)$$

where z (relates to the number of moving charges during the opening and closing of the ion channel), $V_{1/2}$ (corresponds to half activation/inactivation of the channel), A (corresponds to linearity of the activation/inactivation of the channel), and ξ (relates to the asymmetric position of the moving charge in the cell membrane) are fitting parameters and V_m is the membrane potential.

c. Analysis of the recorded ionic currents in voltage-clamp mode at different membrane potentials and importing them into a software for modeling based on the previously listed equations with fitting parameters.

d. Mathematical functions are used to optimize the parameters and to minimize error function. Based on the difference between the recorded and the simulated traces through the whole voltage range, an error function is employed. To find the difference between the fitted curves and the recorded data, the following error functions are used:

$$E_{Max} = \{Abs[R(t_n) - S(t_n)]\} \quad (3.6)$$

where $R(t_n)$ is the recorded signal and $S(t_n)$ is the simulated data at time t_n.

$$\text{Least square: } E_{Lsquare} = \sum_n (R(t_n) - S(t_n))^2 \quad (3.7)$$

$$\text{Weighted least square: } E_{WLsquare} = E_{Lsquare} \quad (3.8)$$

3. Finding parameters to fit recorded APs:

APs are measured with injecting short (2 ms) current pulses in current-clamp mode when they are at resting membrane potential or at a holding potential of -85 mV.

a. Calculation of the membrane potential:

$$\frac{dV}{dt} = \frac{I_{external} - I_{ionic}}{C_M} \quad (3.9)$$

where $I_{external}$ is the injected current to trigger an AP in current-clamp mode.

b. Membrane resistance, resting membrane potential, membrane capacitance, and the amplitude of the injected current are used for modeling. In addition, maximum ionic conductances and leakage current il are calculated from voltage-clamp experiments.

c. The differential equation for the state parameters (m, n, and h) is formed a first-order coupled differential equation system with the membrane potential equation and can be solved with an ordinary differential equation solvers (ODE23).

d. Using a function to find a minimum of a scalar, the parameters are fitted to the experimental data.

Fig. 3.10 shows an excellent fit to the APs using voltage-dependent sodium, potassium, and L-type calcium conductances in the NG108-15 cells. Fig. 3.10A and B shows experimental and simulated APs in control sample and the sample with blocked Na$^+$ channels with 0.5 uM TTX by modifying only the corresponding sodium-channel parameters (see Table 3.1). In addition, Fig. 3.10B−D presents the shape of experimental and fitted APs for both control sample and treated with 0.5 uM tefluthrin (tefluthrin is a channel opener slowing down the inactivation of the sodium channels), respectively.

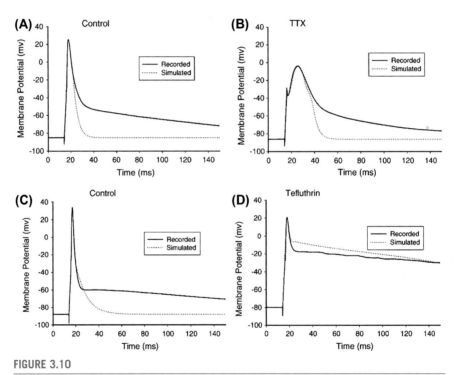

FIGURE 3.10

Fitting simulated action potential (AP) with ion channel parameters and the effects of toxins on the shape of AP. Ion channel parameters are determined from AP recordings based on a computer model of AP generation and parameter fitting. To study the effects of drug, changes of ion channel parameters are investigated [10]. (A,B) Effect of 0.5 uM tetrodotoxin. (C,D) Effect of 0.5 uM tefluthrin. (Solid line) data recorded in current clamp experiments. (Dotted line) results of the simulation using the mathematical model of the NG108-15 cells.

Table 3.1 Average ion channel parameters characteristic of NG-108 cells obtained by parameters fitting to voltage-clamp data ($n = 3$–10) [10].

Average ion channel parameters characteristic of NG108-15 cells obtained by parameter fitting to voltage-clamp data (n = 3–10)

Channel	g	S.E.M	V Rev	S.E.M	Activation								
					z	S.E.M	$V_{1/2}$	S.E.M	ξ	S.E.M	A	S.E.M
Sodium	343.59	183.23	72.35	6.31	5.98	0.30	-46.93	2.46	-0.38	0.01	0.58	0.12
Potassium	25.09	4.81	-80	0	2.78	0.46	-22.52	2.64	-0.26	0.02	2.12	0.16
Calcium	7.45	1.88	32	0	3.15	0.96	-4.67	6.25	-0.3	0.37	0.84	0.37
Leakage	5.22	0.89	-49.40	0.76								
					Inactivation							
					z	S.E.M	$V_{1/2}$	S.E.M	ξ	S.E.M	A	
Sodium					-7.48	1.13	-64.36	4.73	0.41	0.01	1.31	

S.E.M: standard error of mean, VRev: reversal potential.

3.3.5 Photoelectrical stimulation of neuronal cells by an organic semiconductor-electrolyte interface

In this part, an example of using NG108 cells to study the efficacy of a quantum dot (QD) interface is reviewed. Pappas et al. experimentally investigated NG108 cells grown on a QD film consisting of PDDA/HgTe in whole-cell electrophysiology experiments to find the effect of photoactivation of voltage-gated ion channels and firing of APs [17]. The electrophysiological responses after photostimulation with a (532 nm, 800 mW/cm^2) laser have been shown in Fig. 3.11. In the voltage-clamp experiment, using a holding potential of -65 mV (close to resting membrane potential) along with applying a step potential across the membrane leads to an instantaneous inward current (Fig. 3.11A and B) corresponding with the time course of the stimulation of the QD thin film (Fig. 3.11C). This inward current is as a result of cell depolarization and opening and closing of the sodium and calcium channels. In addition, the current response after several repeat of stimulation suggests no oxidative damage or additional cell stimulation as shown in Fig. 3.11D. To measure the functionality of the QD interface, the electrode is placed near the surface and photoinduced potential is measured in current-clamp mode. Fig. 3.11E shows the photoinduced potential at the surface and outside of the cell membrane, which confirms the photoelectrical performance of the interface. In addition, measured membrane leak currents for different hyperpolarizing step voltages under illumination show a linear relation in parallel to that in the dark (see Fig. 3.11F), which indicates that resistance, R, is the same with or without light. This result and the membrane response to different light intensities together show that the QD interface acts as an extrinsic current source that flows across the membrane rather than an intrinsic response that directly activates ion channels.

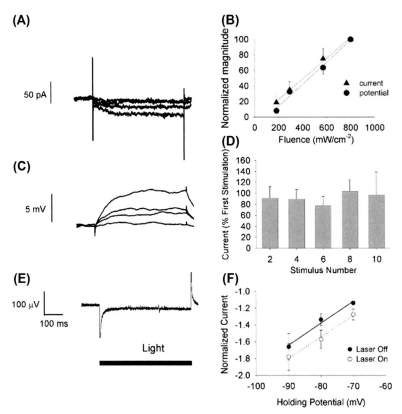

FIGURE 3.11

Photostimulation measurements with whole-cell patch-clamp technique for the cells grown on top of the (PDDA/HgTe) 12PLP films. (A) Cell membrane current measurements in voltage-clamp mode with holding potential of −65 mV at different light fluences. The photostimulation experiment has been performed with a frequency-doubled diode laser at 532 nm for 500 ms at 0.1 Hz in constant wavelength mode. Illumination was at 45 degrees to the NP film surface, and the spot size was approximately 1−1.5 mm in diameter. (B) Membrane potential measured in current-clamp for different light intensity levels. (C) Light-induced potential change from the surface of the NP film. (D) Dependence of photoinduced cell current on pulse number normalized to the current in first pulse. (E) Light-induced potential change in the NP film measured with a micropipette recorded by placing the patch electrode in the bath adjacent to but outside of the cell. (F) Averaged leak currents averaged for three different cells, measured by applying 100 ms voltage steps from −65 to −90 mV, −80 and −70 mV [17].

Also in current-clamp mode, it has been shown that the photostimulus causes depolarization of cells. As Fig. 3.12A and B shows, the mean depolarization level of 2.3 ± 2.4 mV (range = 0−8.6 mV, 28 cells) includes the cells that showed no coupling to the surface at all (<0.5 mV depolarization $n = 8$). In addition, individual photostimulated depolarizations over 10 mV did lead to the generation of

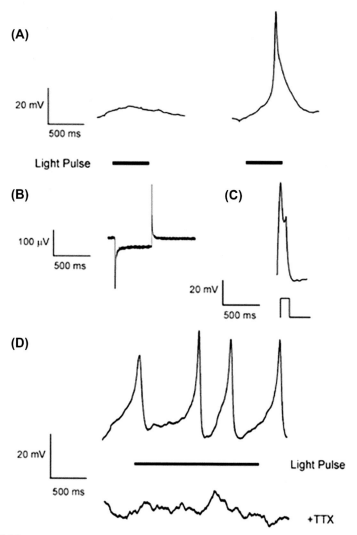

FIGURE 3.12

Current-clamp measurement for recording light-induced action potentials (APs) in NG108 cells grown on LBL HgTe films. (A) Membrane potential changes at subthreshold level (left trace) and AP (right trace) recorded in one NG108 cell. (B) Photovoltage recorded from the QD film surface. (C) Recorded the response of membrane potential by injection of a 0.5 nA, 100 ms current pulse into the cell. This change in membrane potential has similar time course and amplitude to that recorded in photoactivation experiments. (D) Multiple APs were observed after a long (2 s) photostimulus (upper trace) for the NG108 cells grown on (PDDA/HgTe) 12 + (PDDA/clay) 2. In addition, treating with 100 nM TTX inhibited the cells to generate APs (lower trace) [17].

regenerative voltage change characteristic of APs (Fig. 3.12A and D) in 1 of the 28 cells. This shows that the capability of the QD interface to stimulate the cells. However, a more efficient and biocompatible coupling to the QD interface could result in more reliable coupling of the photoactivation and cell stimulation.

3.4 SHSY-5Y cells

SHSY-5Y is a human cell derived from neuroblastoma SK-N-SH. It is differentiated by all-trans-retinoic acid (RA) to obtain more neuron-like features such as neurite outgrowth and axonal developments to mimic neural responses in different studies. Fig. 3.13 shows the optical bright field images from the undifferentiated and differentiated SHSY-5Y cells.

3.4.1 Differentiation protocol for SHSY-5Y cells

There are many reports in the field of neurobiology about the usage of undifferentiated SH-SY5Y cells for human neurons [19−21]. For using SH-SY5Y cells or any other in vitro neuronal system, efficient differentiation of cells into neurons is critical, in order to obtain results that are close to what neurons may response in vivo. Mackenzie et al. reported a protocol that suggests highly viable, homogenous, differentiated neuronal cultures within 18 days for SHSY-5Y cells. Herein, some of the steps are mentioned as following (the tables and instructions for preparing cell growth medias and differentiation medias can be found in the reference) [22]:

   Day 0: Plating Cells for Differentiation
   a. Detachment procedure: At first, attached undifferentiated cells that are already cultured in cultured flasks are washed with 1x PBS, and after aspiration, they are trypsinized with using 1−2 mL warmed 1x 0.05% trypsin-EDTA.

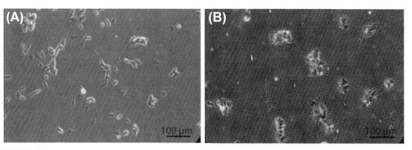

FIGURE 3.13

Optical microscopy images from SHSY-5Y cells for the (A) undifferentiated and (B) differentiated SHSY-5Y cells [18].

b. Incubation for approximately 3 min inside incubator.

c. Adding 10 mL basic growth media to quench the trypsine, gently pipetting and rising the sides of the flask or dish one to three times. Then, they are transferred to a 15 mL conical tube.

d. Centrifuging for 2 min at 1000 x g, and then the old medium containing trypsin is aspirated without disturbing the pellet.

e. With adding new growth media and pipetting for several times, pellet will be resuspended in 5 mL basic growth media.

f. Counting cells by optical microscopy or using a hemocytometer and diluting in basic growth media to 50,000 cells/mL.

g. Seeding 2 mL of cells per 35 mm^2 dish for a total of 100,000 cells per dish and transferring the dishes into incubator.

Day 1: Change Media (Differentiation Media)

a. Preparing aliquot from differentiation media and incubating in a 37°C water bath.

b. Transferring the warmed differentiation media to an incubator (37°C, 5% CO_2) to equilibrate for at least 1 h to reach a proper pH balance prior to use.

c. Adding RA to warmed and equilibrated media immediately prior to add media to dishes. Note: Since RA is sensitive to light, it should be stored in dark bottles at 4°C.

d. Aspirating off old media.

e. Adding 2 mL differentiation media having RA per 35 mm^2 dish and returning to incubator.

Day 3: Refreshing Media (Differentiation Media)

Repeat Section Day 1 (steps a—e)

Day 5: Refreshing Media (Differentiation Media #1)

Repeat Section Day 1 (steps a—e)

3.4.2 Nanoparticle-based plasmonic transduction for modulation of electrically excitable cells

In the following, the study of a micropipette tip coated with gold nanoparticles for both stimulation and inhibition of neural activities in SHSY-5Y cells is reviewed [23]. In a metal layer of gold, photocurrent and heat generated as a result of plasmonic absorption of gold nanostructured coating could lead to cell stimulation. However, it has been found that optoelectrical excitation can lead to cell inhibition. In the optoelectrical excitation, an external current is injected into the electrode, and at the same time, the tip is illuminated with light (at the plasmonic absorption wavelength of 532 nm), therefore an anodic positive extracellular potential produced by the electrode tip coated with gold nanoparticles could result in a negative transmembrane potential and thus inhibit AP generation.

For the photostimulation experiments, a holding current is injected to set the membrane potential to a particular value, and the coated electrode is placed at 2 um distance from the cell membrane. An optical fiber is used to focus light on the electrode and the membrane conductance is recorded using a patch electrode. As Fig. 3.14A shows, after illumination with a 532 nm, 100 mW laser for 10 ms or more, the cell membrane potential is shifted as a function of the cell holding potential. For membrane potential between -30 mV and 0 V, the membrane potential shifts to more positive values under light illumination. However, the amplitude of membrane potential shifts decreases and becomes negative for positive holding potentials. Fig. 3.14A shows an example for time response of the cell membrane to a 10 ms, 100 mW laser pulse. In the figure, for a cell membrane potential of -73.7 mV (applied holding current $= -55$ pA), a positive shift in the evoked membrane potential (*blue curve*) and, for a positive cell potential of $+23.6$ mV (holding current $= +53$ pA), a negative shift (*pink curve*) have been shown. For the photostimulation experiments, APs in response to electrical stimulation are recorded before and after the optical simulation. Fig. 3.14B illustrates the shifts in evoked potential in response to laser nanostimulation and the baseline potential by injecting current for six different cells, each color is related to different cell. Positive shifts in membrane potential indicate partial depolarization, while negative shifts indicate partial hyperpolarization. It has been reported that increasing the laser power and pulse duration leads to higher amplitude of plasmonic-induced membrane potential jumps. The results of photostimulation experiment with a 1 ms, 532 nm green laser pulse with 100 mW power show triggering APs (Fig. 3.14C shows for a representative SH-SY5Y cell when the nanoelectrode was illuminated). Also, optically stimulated AP recordings from six different cells for the range of 75−120 mW of laser power intensity have been illustrated in Fig. 3.14D. For the powers lower than 60 mW, electrically evoked APs prior to and after the optical stimulation in response to electrical current pulses are the same. Control experiments without nanoeletrode but with laser illumination (-10 ms, 100 mW) also confirms the functionality of the electrode and no evoked potential by using only light could be obtained.

Inhibitory cell responses: Whole-cell current-clamp is also used for studying cell inhibition similar to cell stimulation experiments. The nanostructured gold electrode is illuminated by laser pulses delivered using the optical fiber. Current pulses are injected into the cell via patch pipette. With injecting 180 pA, 300 ms current pulses, APs are recorded in the SHSY-5Y cells. With injecting electrical current to the electrode at the same time as illumination (300 ms, 120 mW), decreases in the amplitude and rate of the APs could be observed as compared with the control conditions, i.e., electrical stimulation alone. Fig. 3.15A shows recorded APs before and after the optical stimulation experiments, and no deleterious effects are detected. With increasing laser power, the inhibition becomes more significant (see Fig. 3.15B). The most significant inhibition is seen when the laser pulse led the electric current pulse by a few milliseconds (5−15 ms) (see Fig. 3.15C). To find the mechanism behind the inhibition from analyzing the shape of APs, it is important to remember that in AP generation, depolarization happens predominantly because of Na^+ influx and repolarization happens mainly because of K^+ efflux.

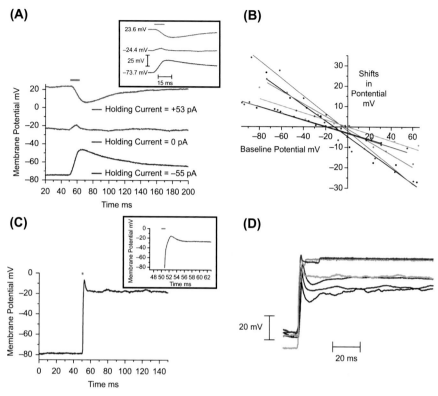

FIGURE 3.14

Whole-cell current-clamp recording for measuring the photoactivation of SHSY-5Y cells grown on top of the photoactive layer. (A) Time response in cell membrane potential when illuminating with a 532 nm green laser (denoted by the *green bar*), 10 ms pulse, having a power of 100 mW shined on the tip of the nanoelectrode. The inset shows the onset response in more detail (faster time scale). The figure shows shift in evoked potentials at three different baseline potentials of −73.7 mV (applied holding current = −55 pA), −24.4 mV (applied holding current = 0 pA), and +23.6 mV (applied holding current +53 pA). (B) The change in evoked photoinduced potential varied with different fixed membrane potential by injecting constant injecting currents: positive for −30 mV or less and negative for 20 mV or more. Shifts in evoked potential versus baseline membrane potential for six different cells are shown, each represented by a different color. (C) A representative action potential (AP) recorded from an SH-SY5Y cell under illumination with a 1 ms laser pulse at 100 mW laser power. The inset shows the zoomed portion of the onset response of the same. The cell membrane baseline potential was −78.5 mV (applied holding current = −35 pA). (D) Evoked APs recorded from the six different SHSY-5Y cells have been shown in different colors [23].

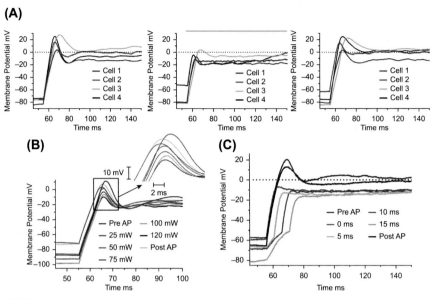

FIGURE 3.15

Inhibition of action potentials (APs) of SH-SY5Y neuron-like cells with current-clamp recording. When laser illumination and electric current pulses were applied at the same time, a reduction in magnitude of action potentials (APs) was observed. (A) The time responses for four different cells. Left: cells stimulated with electric current pulses shows APs (pre-AP) with injecting electrical pulse current (180 pA, 300 ms)—control conditions. Middle: a 300 ms, 120 mW, 532 nm green laser pulse (shown by *green bar*) was illuminated at the same time with injecting an electric current pulse (180 pA, 300 ms); the response of the membrane potential shows a reduction in the amplitude of AP. Right: postlight experiments, which show that the membrane potential returns to the original value. The increase of 20K in temperature around was experienced by ~5–10% cell area. (B) The inhibition of AP is affected by laser power (with the pulse width of 300 ms). By increasing laser power, inhibition becomes more significant—AP peak decreases with laser power. (C) The inhibition of AP is more significant when the laser pulse is applied a few milliseconds sooner than the electrical pulse as shown in a representative cell [23].

Since Na^+ influx leads to depolarization and K^+ efflux causes repolarization, it is expected that during inhibition, the degree of K^+ efflux contributes more than the rate of Na^+ influx. To understand the mechanism of inhibition, depolarization and repolarization phases of the APs are analyzed as shown in Fig. 3.16A. In this figure, a different part of AP has been analyzed on the following terms: (1) peaks (AP peaks), (2) rising slope of the AP (rate of depolarization), (3) normalized rising slope to the initial value of the rising slope for electrically produced APs, (4) amplitude from the initial peak value and the first minimum of the AP (Fig. 3.16E), (5) first

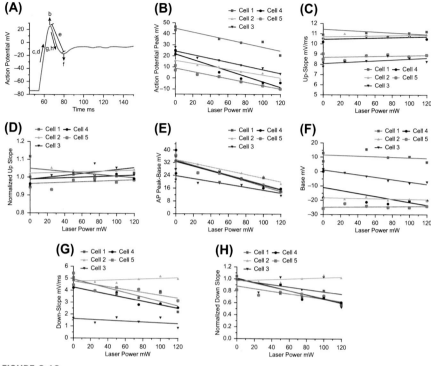

FIGURE 3.16

Analysis the shape of action potential (AP) to investigate inhibition mechanism. (A) A representative figure indicating different phases of AP for the various analysis done for inhibition experiments as function of laser power, shown in subsequent parts. (B) Increasing the laser power decreases AP peak, shown for five different cells. (C) Increasing the laser power does not affect the rate of rise of AP (mV/ms) as shown in upslope versus laser power curves for five different cells. (D) Normalized rising slope values with initial upslope (pre-AP upslope), which shows same trend of having no change with laser power like absolute values in (C). (E) The amplitude of AP peak from the first minima after peak decreasing with laser, as a result of the decrease in peak values of AP (B). (F) The first minima remains similar, and it does not vary with laser power. (G and H) The falling slope and its normalized version with pre-AP decrease with laser power. The electrical stimulations alone as control condition have been shown by points with zero laser power [23].

minimum after peak value (Fig. 3.16F), and (6) the falling slope of the AP (rate of repolarization) and normalized downslope of AP (similar to rising slopes, down-slopes were also normalized with initial value of downslope for electrical APs without optical stimulation experiment). With increasing laser power, the AP peak value decreases (see Fig. 3.16B). In lower power ranges, the rising slope (absolute and normalized) depolarization rate has been shown to remain unaffected

(Fig. 3.16C and D), while the falling slope (absolute and normalized) repolarization rate decreases by increasing laser power (see Fig. 3.16G and H). The first minimum after the AP peak, i.e., base value does not change by increasing laser power (see Fig. 3.16F), and the amplitude from the peak to base values decreases by laser power (Fig. 3.16E).

3.5 Neuro-2a

Neuro-2a cells are a mouse neuroblastoma cell line that has been widely used for neural differentiation, studying neurotoxicity, and neural stimulation. Fig. 3.17 shows optical microscopy images form undifferentiated and differentiated Neuro-2a cells after 4, 72, and 96 h treatment with RA. Cell differentiation often causes decrease of cell growth, or in other words, it alters the rate of macromolecule synthesis and degradation. On the other hands, autophagy is a central controller of cell growth and differentiation. Zeng et al., with analysis of the abundance of LC3-II or LC3-II/LC3-I by immunoblot analysis, presented the evidence that autophagy plays a role in a process for bulk degradation of cytoplasm activated during RA-induced neuronal differentiation of neuroblastoma Neuro-2a cells [24]. Fig. 3.17 shows the Neuro-2a cells before and after differentiation with 20 μM RA for different treatment times of 4, 72, and 96 h. For longer treatment times,

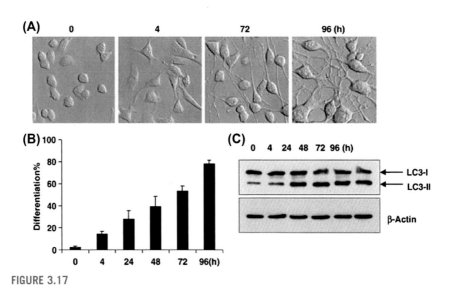

FIGURE 3.17

Morphological changes of Neuro-2a cells before and after differentiation with 20 μM retinoic acid. (A) Neuro-2a cells before treatment for 4, 72, and 96 h. (B) Percentage of differentiation for different treatment times. (C) Immunoblot analysis of cells with anti-LC3 and anti-β-actin to evaluate the level of autophagy [24].

the percentage of differentiation increases (see Fig. 3.17B) up to more than 80% after 96 h. Fig. 3.17C also shows lysate cells were immunoblotted with anti-LC3 and anti-β-Actin, which reveals that after the induction of differentiation, both LC3-II and the ratio of LC3-II/LC3-I were significantly increased.

3.5.1 Differentiation protocol for Neuro-2a cells

The protocol for the neural differentiation has been reported as (1) seeding cells at a density of 100 cells/mm^2 and after overnight incubation, (2) treatment with 20 μM RA (Sigma) in DEME plus 2% FBS for 4, 24, 48, 72, or 96 h [25], and (3) refreshing medium for every 24 h is recommended.

3.5.2 Excitability of Neuro-2a cells

Electrical excitability of Neuro-2a cells based on their neurophysiological behaviors has been reported, and it has been shown that this cells have sodium channels and are able fire APs after neural differentiation [26,27]. In addition, the Neuro-2a cells were used as a model system for photoelectrical stimulation purpose by monitoring of sodium current [28].

3.5.3 Photoelectrical stimulation of neuronal cells by an organic semiconductor-electrolyte interface

Oliya et al. used Neuro-2a cells for studying photostimulation experiments for an organic-electrolyte interface [28]. They used 2, 4-bis [4-(N, Ndiisobutylamino)-2, 6-dihydroxyphenyl] squaraine, shortly named SQIB, as a photovoltaic layer to stimulate Neuro-2a cells by photoelectrical mechanism. Probed functionality of Neuro-2a cells has been shown in Fig. 3.18 by electrophysiological patch-clamp recordings. As it shows in Fig. 3.18A, fast sodium inward currents were recorded by applying depolarizing voltage steps in voltage-clamp (Fig. 3.18B), and inset in Fig. 3.18B shows also inward sodium currents are followed by slower potassium outward currents. These recordings provide evidence of the functionality of Neuro-2a cells grown on the photoactive layer.

Photostimulation experiments: The response of cells in voltage-clamp and current-clamp mode recorded from Neuro-2a cells under photoelectrical stimulation of cells has been shown in Fig. 3.18. Charge accumulation at the interface of photoactive layer and electrolyte interface has been proposed to be the mechanism of the photostimulation as also illustrated in Fig. 3.18C (which illustrates the schematic of the photoactive interface and experimental setup for photostimulation of Neuro-2a cells). As it shows, the Neuro-2a cell has a resting membrane potential of $V_M = -75$ mV, while during photostimulation at 690 nm, it shows capacitive transient responses of two rapid depolarizations in the order of $\Delta V_M = 0.4$ mV (see Fig. 3.18D). When the light turns on, the charge accumulation at the photoactive layer causes positive ions to drift within the electrolyte to the interface forming a

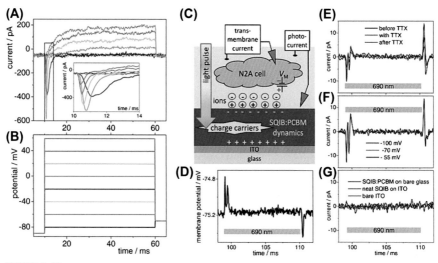

FIGURE 3.18

Whole-cell patch-clamp recordings of Neuro-2a cells grown on top of the photoactive layer and schematic of the photostimulation mechanism by the photoinduced charge accumulation in the photoactive layer. (A) Measured membrane current showed ionic inward (sodium) and outward (potassium) currents, respectively, for various depolarizing voltage pulses (B). The inset in (A) shows the voltage pulses applied across cell membrane. (C) Schematic of the experimental setup describing the mechanism of charge generation upon light illumination, which produces transient photocurrent and consequently cell depolarization. (D) Photoelectrically induced membrane potential of Neuro-2a cell stimulated by the transient photocurrent, recorded in current-clamp mode. (E) Recorded whole-cell membrane currents normalized to 10 pF of Neuro-2a cell under photostimulation: before and after treatment with TTX. (G) Control experiments by variation of interface architecture. The effect of Illumination was through the Ringers's solution with 10 ms pulses centered at 690 nm in all cases [28].

Helmholtz double layer. Consequently, the extracellular environment above the photoactive layer becomes negative and leads to increase of intracellular potential probed as depolarization. On the other hands, when the light turns off, recombination of charges in the photoactive layer leads to the drift of ions within the electrolyte in reverse direction, causing in a rapid hyperpolarization of the cell membrane. The response of membrane current also indicates capacitive currents with a similar signal to the photoelectrically induced membrane potential, and it indicates the passive response of the cell membrane. This capacitive current can be interpreted based on the Hodgkin-Huxley model: $Ic_M = C_M \, (dV_M/dt)$ in which V_M is the membrane potential and C_M is the membrane capacitance. In addition, the photoactive layer producing similar photocurrent and such a capacitive membrane current is as a result of passive response of the cell. Since this capacitive stimulation has two polarities, it can depolarize and hyperpolarize the cells.

This photoinduced change in the membrane potential is not likely to cross the threshold for the opening of voltage-gated sodium channels. Although, the measured ionic transmembrane currents in the patch-clamp recordings would be distinguishable (see Fig. 3.18E and F). Based on a set of intracellular depolarizing voltage pulses (see Fig. 3.18B), an increase of -20 mV in membrane potential V_M or change in membrane potential ΔV_M of a few tens of millivolts is needed.

3.6 Conclusion

In this chapter, basic information about different neuron-like cells including protocol of differentiation and their electrophysiological characteristic were explained. As it has been described, based on the mathematical analysis of the AP shape and recorded ionic currents, the mechanism of cell stimulation and inhibition can be examined. Moreover, some studies about photostimulation of cells with QDs and gold nanostructured electrode have been reviewed as examples for the usage of this neuron-like cells in neural stimulation experiments.

References

[1] Rudy B, et al. Nerve growth factor increases the number of functional Na channels and induces TTX-resistant Na channels in PC12 pheochromocytoma cells. Journal of Neuroscience 1987;7(6):1613−25.

[2] Drubin DG, et al. Nerve growth factor-induced neurite outgrowth in PC12 cells involves the coordinate induction of microtubule assembly and assembly-promoting factors. The Journal of Cell Biology 1985;101(5):1799−807.

[3] Dichter MA, Tischler AS. Nerve growth factor-induced increase in electrical excitability and acetylcholine sensitivity of a rat pheochromocytoma cell line. Nature 1977;268:501−4.

[4] Barchi RL. Biochemical studies of the excitable membrane sodium channel. In: International review of neurobiology. Elsevier; 1982. p. 69−101.

[5] Bahmani Jalali H, et al. Effective neural photostimulation using indium-based type-II quantum dots. ACS Nano 2018;12(8):8104−14.

[6] Mobley WC, Schenker A, Shooter EM. Characterization and isolation of proteolytically modified nerve growth factor. Biochemistry 1976;15(25):5543−52.

[7] Letourneau PC. Cell-to-substratum adhesion and guidance of axonal elongation. Developmental Biology 1975;44(1):92−101.

[8] Thakore V, Molnar P, Hickman JJ. An optimization-based study of equivalent circuit models for representing recordings at the neuron−electrode interface. IEEE Transactions on Biomedical Engineering 2012;59(8):2338−47.

[9] Gheith MK, et al. Stimulation of neural cells by lateral currents in conductive layer-by-layer films of single-walled carbon nanotubes. Advanced Materials 2006;18(22):2975−9.

[10] Mohan DK, Molnar P, Hickman JJ. Toxin detection based on action potential shape analysis using a realistic mathematical model of differentiated NG108-15 cells. Biosensors and Bioelectronics 2006;21(9):1804−11.

[11] Whitemarsh RC, et al. Model for studying Clostridium botulinum neurotoxin using differentiated motor neuron-like NG108-15 cells. Biochemical and Biophysical Research Communications 2012;427(2):426−30.

[12] Liu J, et al. Voltage-gated sodium channel expression and action potential generation in differentiated NG108-15 cells. BMC Neuroscience 2012;13(1):129.

[13] Teorell T. Excitability phenomena in artificial membranes. Biophysical Journal 1962; 2(2):25−51.

[14] Szlavik RB, Jenkins F. Varying the time delay of an action potential elicited with a neural-electronic stimulator. In: The 26th annual international conference of the IEEE engineering in medicine and biology society. IEEE; 2004.

[15] Hodgkin AL, Huxley AF. A quantitative description of membrane current and its application to conduction and excitation in nerve. The Journal of Physiology 1952;117(4): 500.

[16] Destexhe A, Huguenard JR. Nonlinear thermodynamic models of voltage-dependent currents. Journal of Computational Neuroscience 2000;9(3):259−70.

[17] Pappas TC, et al. Nanoscale engineering of a cellular interface with semiconductor nanoparticle films for photoelectric stimulation of neurons. Nano Letters 2007;7(2): 513−9.

[18] Cheung Y-T, et al. Effects of all-trans-retinoic acid on human SH-SY5Y neuroblastoma as in vitro model in neurotoxicity research. Neurotoxicology 2009;30(1):127−35.

[19] Yun S-I, et al. A molecularly cloned, live-attenuated Japanese encephalitis vaccine SA14-14-2 virus: a conserved single amino acid in the *ij* hairpin of the viral E glycoprotein determines neurovirulence in mice. PLoS Pathogens 2014;10(7):e1004290.

[20] Kalia M, et al. Japanese encephalitis virus infects neuronal cells through a clathrin-independent endocytic mechanism. Journal of Virology 2013;87(1):148−62.

[21] Xu K, et al. Replication-defective HSV-1 effectively targets trigeminal ganglion and inhibits viral pathopoiesis by mediating interferon gamma expression in SH-SY5Y cells. Journal of Molecular Neuroscience 2014;53(1):78−86.

[22] Shipley MM, Mangold CA, Szpara ML. Differentiation of the SH-SY5Y human neuroblastoma cell line. Journal of Visualized Experiments 2016;(108):e53193.

[23] Bazard P, et al. Nanoparticle-based plasmonic transduction for modulation of electrically excitable cells. Scientific Reports 2017;7(1):7803.

[24] Zeng M, Zhou J-N. Roles of autophagy and mTOR signaling in neuronal differentiation of mouse neuroblastoma cells. Cellular Signalling 2008;20(4):659−65.

[25] Gallo R, et al. REN: a novel, developmentally regulated gene that promotes neural cell differentiation. The Journal of Cell Biology 2002;158(4):731−40.

[26] Combs DJ, et al. Tuning voltage-gated channel activity and cellular excitability with a sphingomyelinase. The Journal of General Physiology 2013. https://doi.org/10.1085/jgp.201310986.

[27] Chao C-C, et al. HMJ-53A accelerates slow inactivation gating of voltage-gated K^+ channels in mouse neuroblastoma N2A cells. Neuropharmacology 2008;54(7): 1128−35.

[28] Abdullaeva OS, et al. Photoelectrical stimulation of neuronal cells by an organic semiconductor−electrolyte interface. Langmuir 2016;32(33):8533−42.

In vivo electrophysiology

4

4.1 Introduction

Extracellular in vivo measurements of neurons enabled us to understand the functionalities of different parts of the brain and to *read neural circuits.* However, based on the in vitro single whole-cell patch-clamp electrophysiology, the electrophysiological behaviors of the single cells and their interactions with different neural interfaces could be investigated. To record neural activities in extracellular medium, electrodes are implanted in the brain tissue in the space between neurons. In general, extracellular ionic current flows in the extracellular medium cause potential drop at the electrode tip which is called local field potentials (LFPs). Fig. 4.1 illustrates different stages of neural data processing for in vivo electrophysiology setup. In this setup, a lowpass filter is used to reduce high-frequency noises and visualize LFPs from surrounding the electrode (around 1 mm in diameter). The source of LFPs is the input currents generated by the dendrites of the surrounding neurons. In fact, when there are many cell bodies that are (partially) aligned, the intensity of LFPs will be high due to the creation of coherent dipoles in the extracellular medium [1]. Bandpass filtering selects signals coming from a few neurons close enough to the electrode plus background activity produced by neurons (black trace in the bottom panel of Fig. 4.1).

After the bandpass filter, the output signal contains superimposition from activity of different neurons. The next step is to determine which spike corresponds to which neuron. In general, each neuron produces signals that have a particular shape in a given electrode. This unique characteristic is because of the difference in the morphology of dendritic tree in different neurons, ionic conductances and the distribution of ion channels, the physical distance and direction relative to the recording electrode, and the properties of the extracellular medium [2]. Thanks to spike-sorting detection algorithms, spikes from the recorded signal are classified into different clusters based on their shapes [3]. Clustered and separated spiking events are related to different neural sources, and those with a low signal-to-noise ratio are associated with multiunit activity (Fig. 4.1). The multiunit cluster is characterized by a relatively low amplitude and not following the refractory period of single neurons—i.e., the time period of spikes is within less than 2.5 ms [4]. Based on extracellular recordings and spike-sorting detection algorithms, the activity of a

Electrophysiology Measurements for Studying Neural Interfaces. https://doi.org/10.1016/B978-0-12-817070-0.00004-X

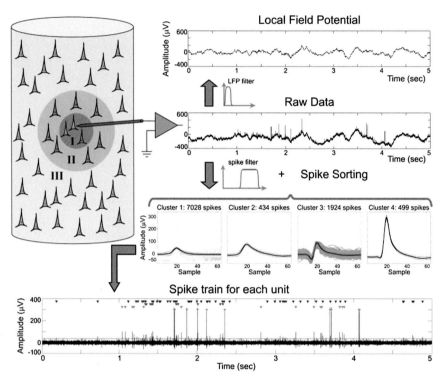

FIGURE 4.1

Extracellular recordings and the spike-sorting procedure. An extracellular recording (raw data) from the human right entorhinal cortex. The low-frequency component of the recorded signal is related to the local field potential (between 1 and 100 Hz in this example). In the higher-frequency component (between 300 and 3000 Hz in this example; black trace in bottom panel), a superposition of several effects can be found. Neurons in zone III (more than ~ 140 μm away from the tip of the electrode) contribute to the background noise; therefore, their spikes cannot be detected. The neurons in zone II produce spikes with higher amplitude in comparison with the background noise, but they cannot be classified into different units, thus being related to the multiunit activity (cluster 1). Finally, the spikes produced from neurons in zone I (less than ~ 50 μm away from the tip of the electrode) have even higher amplitudes, and sorting algorithms enable us to label the recorded spikes to the different neurons that generated them (clusters 2–4), hence the so-called single-unit activity. Spike train includes the sequence of spikes associated with each cluster. The bottom panel shows the spike events, each marked with a triangle color coded according to the isolated clusters [1].

few neurons per electrode can be recorded for a few hours and in the case of acute recordings in each recording session, up to months or even years with stable implanted electrodes [5].

4.2 Spike sorting
4.2.1 Current spike-sorting strategies
Fig. 4.2 shows the basic steps for spike-sorting procedure. The first step filtering the recorded raw data by a bandpass filter to detect spikes. In general, the bandpass filter also acts as a comparator for the spike detection from the filtered data. The next step is feature analysis of the spike shapes based on amplitude, width, and energy of the spikes, which provides input to the clustering stage, which classifies the activity of the different neurons in the feature space and associate each cluster to a unit. In the following, each step is explained in more detail.

4.2.2 Filtering
As briefly mentioned, recorded raw data from extracellular measurements containing the LFPs and each signal can be extracted by filter stage (spanning a range of frequencies from low frequencies up to several kHz), respectively. The filter stage includes a bandpass filter that defines an amplitude threshold, and generally, a second digital filter with the bandwidth of 2700 Hz (300 and 3000 Hz) is used.

4.2.3 Detection
Filtered raw data including clear spikes and a background noisy activity can be detected by using an amplitude detector adjusted with a threshold level. For very low threshold levels, input noises can cause false positive events, and for very large threshold levels, low-amplitude spikes are neglected. As selecting proper threshold values depends on the signal-to-noise ratio of the recording signal, it is more logical to use an automatic threshold approach that examines the standard deviation of the noise, i.e., threshold $= k * \hat{\sigma}n$, where k is a constant typically between 3 and 5.

4.2.4 Feature extraction
The aim of feature extraction is to facilitate the classification and bypass the signals that are just random variations. The simplest spike-sorting algorithms extract features based on the amplitude of the spikes. However, generated spikes from the nearby neurons can have similar peak amplitude but with a different shape. One solution is using "window discriminators," which apply one or more time-

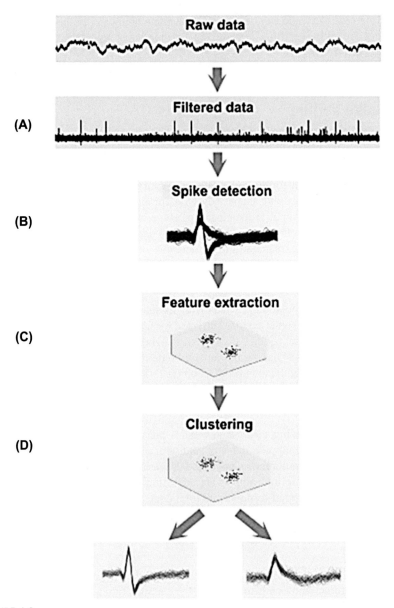

FIGURE 4.2

Fundamental steps in spike-sorting procedure. Spikes are recorded in a raw data format; (A) a bandpass filter is used with a bandwidth between 300 and 3000 Hz, which allows the most useful important information for spike sorting to pass. Next, (B) detection of spikes through using an amplitude threshold applied to the filtered data. (C) Extraction of spike features, achieving a dimensionality reduction. Finally, those extracted features are amplitude windows that are defined, and the waveforms crossing them are assigned to a particular unit [3]. Although this method also demands manual intervention and it is not possible practically, large datasets are recorded from large numbers of electrodes. In addition, readjustment of the windows will be required during an experiment, as the recordings are nonstationary and consequent changes in the spike shapes occur.

Moreover, considering that the number of recorded spikes is large, for instance, 80,000 during an hour-long experiment, and spike signals consist of 48 data points (based on the sampling rate in the data acquisition system monitors a raw voltage trace), we need to solve this problem (separating of the spikes) in a 48-dimensional space. In this regard, principal component analysis (PCA) enables us to reduce this dimension by compressing the data [7−9]. PCA is a linear transformation of variables that creates a new dataset, which is a weighted linear combination from an input dataset, in a new space with fewer dimensions. For that, PCA keeps as much of the variance in the dataset (i.e., the separation between points) as possible while it gets rid of correlations (i.e., less redundancy). Fig. 4.3A and B shows superimposed demeaned spikes and their obtained PCA components, respectively. It is important to note that in real recordings, there are many units contributing to multiunit activity, so there is considerable noise and large number of spikes. Moreover, each spike waveform of a neuron can vary over time during the firing. For example, as shown in Fig. 4.3C, it is possible that the amplitude and shape of the spikes from a single neuron can change, especially at higher activity levels. Recorded spikes can be written as rows of the matrix S. This matrix represents a weighted sum of orthonormal eigenvectors [11]:

$$s = \sum_{n=1}^{N} \lambda_n u_n \tag{4.1}$$

The features, or PCA scores, are computed as

$$s_i = u_i^T s \quad i = 1, 2, 3, \ldots, d. \tag{4.2}$$

In most cases, $d = 2$ features (sometimes $d = 3$ depending on how separable the clusters are) to represent each spike. Subsequently, algorithms such as expectation-maximization (EM) [12] or fuzzy c-means [13] are used for clustering the features.

4.2.5 Clustering

Finally, the feature space containing PCA components is partitioned into distinct clusters. In this step, the points are labeled to different clusters, and each cluster

used for clustering step; (D) classification of the waveforms and association of each cluster to an unit.

Adapted from Quiroga RQ, Nadasdy Z, Ben-Shaul Y. Unsupervised spike detection and sorting with wavelets and superparamagnetic clustering. Neural Computation. 2004;16(8):1661−1687.

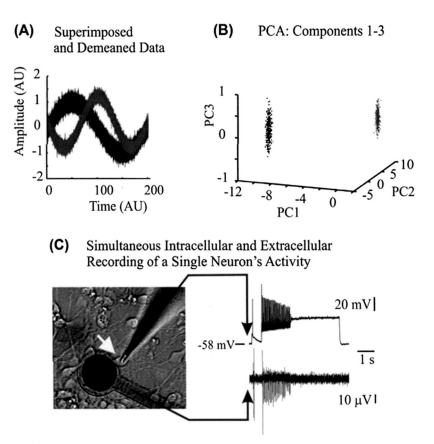

(A)   Superimposed and Demeaned Data

(B)   PCA: Components 1-3

(C)   Simultaneous Intracellular and Extracellular Recording of a Single Neuron's Activity

FIGURE 4.3

Principal component analysis (PCA) for spike sorting. (A) Superimposed demeaned noisy measurements of two overlapping simulated waveforms. Both amplitude and time are in arbitrary units (AU). (B) The projection of the spike measurements on the first three principal components (PC1, PC2, and PC3). It shows that while the two types of waveform (red and black dots) are distinguished by the first two components, the third, vertical component (PC3) does not contribute to separate the two types. Panels (A and B) are generated by MATLAB script pr28_2. (C) An example of intracellular (top trace) and extracellular (bottom trace) measurements from a cultured hippocampal pyramidal cell (white arrow) at the same time. It illustrates that at higher levels of neuronal depolarization, the shape of spikes (amplitude) varies, which can lead spike sorting to fail. The current-clamp measurement shows the action potential generation after a depolarizing current injection [10]This figure is adopted from Wim van Drongelen, Chapter 28 - Decomposition of Multichannel Data in Signal Processing for Neuroscientists (Second Edition) 2018, Pages 579-617, Experimental data in panel C from Dr. A.K. Tryba, with permission.

is related to a different neuron. This can be manually realized by finding boundaries based on similarities among the patterns [14]. However, it can be time-consuming and inaccurate (errors that arise from both the limited dimensionality of the cluster cutting space and human biases [8,15]). This step is very challenging, as there could be a large variety of shapes for clusters and it is not defined how many there are beforehand. The final step is labeling the waveforms in each cluster as belonging to a single neuron (or unit). Fig. 4.4A shows PCA feature space plots for a dataset, where PC1 and PC2 are the first and second principal components, respectively. For this plot, there can be several clusters with different degrees of cluster separability. Based on an unsupervised EM algorithm, five distinct spike shapes that correspond to five putative single units could be found (as shown in Fig. 4.4B).

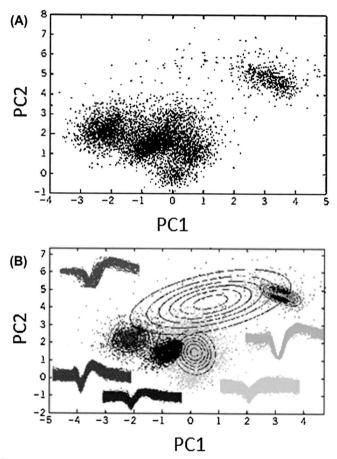

FIGURE 4.4

Feature extraction from PCA components and EM unsupervised cluster cutting of 6469 spike events. (A) Projection of unlabeled spikes onto the two dominant eigenvectors of the data matrix. (B) Five unique units can be separated with this approach as indicated by the EM contour plots. It is possible to extract more units with user supervision—for instance, through breaking the lightest gray cluster into two or three smaller clusters [11].

4.3 Animal studies

4.3.1 In vivo study of retinal prosthesis (closed-loop photoelectrical stimulation)

Photoelectrical stimulation as a novel technology has enabled retinal prosthesis in a wireless and non-invasive manner. Keith et al. developed a subretinal photovoltaic platform consisting of arrays of photodiodes to stimulate the inner retinal neurons through patterns of photocurrents. Fig. 4.5 shows a hexagonal array of pixelated photodiodes for the aim of photoelectrical retinal prostheses. As described in Chapter 2, based on the photoelectrical stimulation technology, generated photocurrents by the photovoltaic interfaces can stimulate the neurons. In each pixel, several photodiodes are connected in a series and generate faradaic photocurrents under illumination (Fig. 4.5C). To localize and more efficiently discharge the photocurrent, a return electrode and a shunt resistance have been embedded into each pixel, respectively. Fig. 4.D shows the position of a healthy rat retina which is sandwiched between the photovoltaic array (PVA) to stimulate and MEA to record from hundreds of retinal ganglion cells (RGCs). To measure the efficacy of the retinal prothesis, the

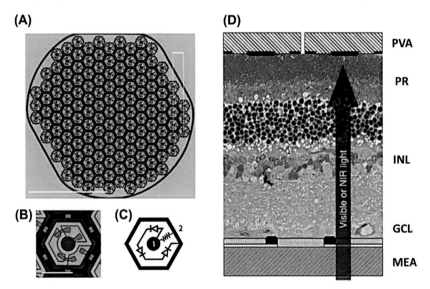

(A) (B) (C) (D)

PVA

PR

INL

Visible or NIR light

GCL

MEA

FIGURE 4.5

Photovoltaic platform and in vitro experimental setup. (A) The platform consists of 70 μm pixels separated by 5 μm trenches arranged in a 1 mm-wide hexagonal pattern, with the adjacent rows separated by 65 μm. Scale bar, top right-hand corner: 65 μm; bottom left-hand corner: 500 μm. (B and C) Each pixel consists of two to three (shown here) photodiodes connected in series around a centered active area (1) and surrounding area is the return electrode (2). Scale bar: 50 μm. (D) Schematic illustrating of a healthy rat retina placed between a transparent multielectrode array (MEA) and the photovoltaic array (PVA). Visible light stimulates the photoreceptors (PR), while NIR (880–915 nm) illumination generates biphasic pulses of current in the photovoltaic pixels, which stimulate the cells in the inner nuclear layer (INL) [16].

effect of photostimulation on animals with retinal degeneration (RCS) with the implant and wild-type (WT) animals is studied. Then, photoelectrically induced eVEP from the RCS animals with the implant and light-induced eVEP from WT animals are recorded. In principal, the photostimulation thresholds (light intensities) of eVEPs for both groups are compared to investigate the effect of the implant.

In in vivo experiments, the recorded visually evoked potentials (VEPs) responded to either the implant activated to near IR light (eVEP) or normal retina to light are analyzed (Fig. 4.6). A camera is installed in front of a slit lamp to monitor the stimulus on the retina and at the same time to properly align the laser beam with the implant. Recording eVEP signals upon photoelectrical stimulation by the subretinal implant shows that visual information has been transferred by the implant in the retina and received at the visual cortex. Basically, different light signal properties such as light intensity and pulse width can modulate cortical responses similar to modulation of the retinal ganglion cells (RGC) responses in vitro.

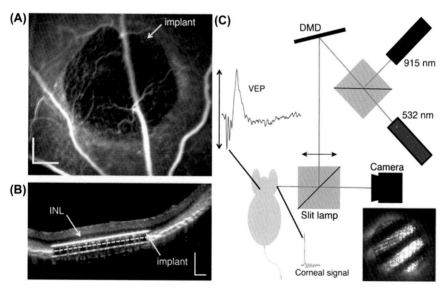

FIGURE 4.6

In vivo subretinal implantation and stimulation setup. (A) Fluorescein angiography 1 week after surgery demonstrates normal retinal blood perfusion above the implant with no leakage. Scale bar: 200 μm. (B) Optical coherence tomography (OCT) image shows that the inner layer of retina has been preserved, with the inner nuclear layer (INL) located approximately 20 μm above the upper surface of the implant (white line). The 30 μm implant appears thicker due to its high refractive index. The yellow dashed line illustrates the actual position of the back side of the implant on top of the retinal pigmented epithelium (RPE). Scale bar: 200 μm. (C) Stimulation setup for VEP recordings. Illuminated visible (532 nm) and NIR (915 nm) light from lasers, which generates the spatial patterns, is projected onto retina with a digital micromirror device (DMD), as shown in the photograph insert. The transcranial electrodes are used to record VEP signals of cortical activity at the same time as the corneal potential, which shows the stimulation pulses from the implant [16].

Fig. 4.7 shows the result of photostimulation with and without implants. As it shows, the amplitude of the eVEP increases with the light intensity levels from 0.125 mW/mm^2 and 1 mW/mm^2, and it saturates at higher light intensity levels (Fig. 4.7A). The cortical response also shows increase by pulse duration from 1 ms to 10 ms, which saturates at longer pulse durations (Fig. 4.7B). This result

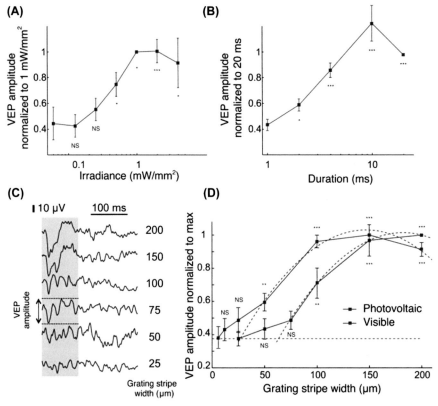

FIGURE 4.7

(A) In vivo photostimulation to the study the effect of light intensity modulation on visual acuity VEP and (B) for the pulse width under full-field illumination (n = 9 for WT animals and n = 7 for RCS animals). (C) Sample VEP traces show the effect of different grating stripe widths. The VEP amplitude has been defined as the peak-to-peak variation of the signal during the first 100 ms following grating alternation (gray-shaded area) for prosthetic stimulation. Visible light triggered slower, and longer-lasting responses and the amplitude have been measured during the first 300 ms after alternation. Responses show to decrease to the noise level with 50 μm stripes. (D) VEP amplitude for visible gratings (blue) and prosthetic stimulation (red) decreases with reducing width of the stripes. Based on the crossing point of the parabolic fits with the noise level (dashed lines), the acuity limit and its associated uncertainty are estimated, which corresponds to 27 ± 9 μm/stripe for visible light and 64 ± 11 μm/stripe for prosthetic stimulation (n = 7 WT animals with visible light and n = 7 RCS animals with prosthetic stimulation). Error bars show standard error of the mean. *NS*, not significant; *, $P < .05$; **, $P < .01$; ***, $P < .001$, one-tailed Welch t-test ,performed against the lowest irradiance, duration, and grating size groups in (A,B and D), respectively [16].

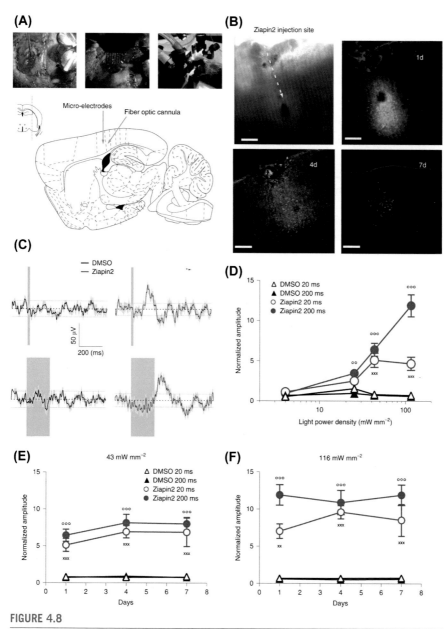

FIGURE 4.8

In vivo prosthetic photostimulation and visual acuity measurements. (A), Schematic representing the stereotaxic injection of Ziapin2 (200 μM in 1 μL 10% DMSO) in the somatosensory cortex (S1ShNc, 2 mm anterior to lambda, 2 mm lateral to midline, and −723 μm ventral to brain surface) and the 16-microelectrode array implant for LFP recordings coupled with optical fiber for photostimulation. (B), Bright field image (left) and indicates that the cortical response to photovoltaic subretinal stimulation (for both light intensity and duration) is over an order of magnitude wide dynamic range. Photostimulation thresholds of VEPs were shown to be similar for both WT and RCS rats, with 10 ms pulses and 2 Hz repetition rate, and with the stimulation threshold level of 0.55 ± 0.08 mW/mm^2.

4.3.2 In vivo study of retinal prosthesis (photocapacitive stimulation)

As described in Chapter 2, photocapacitive stimulation of cells could induce photo-capacitive displacements in the membrane potential by a photoswitch (Ziapin2) in the cell membrane. Fig. 4.8A shows the in vivo setup to monitor the effect of injected photoswitch particles in the somatosensory cortex of mice via an implanted micro-electrode array with an optical fiber. Photostimulation experiments for different time periods of shortly after surgery (30−60 min) and 1, 4, and 7 days after the Ziapin2 injection show a diffusion diameter in the range of 1 mm and persisted up to 7 days from the time of injection (Fig. 4.8C). As shown in Fig. 4.8C, photostimulation for different light power intensities could trigger cortical activity measured through LFPs that peaked at about 200 ms after light onset (Fig. 4.8C). Also, the results show that the response of LFP amplitude with respect to vehicle-injected animals is more significant for 200 ms stimuli (Fig. 4.8D). Based on the recorded light-evoked LFP responses, efficient photostimulation of cortical activity has been recorded up to 7 days after injection (Fig. 4.8E and F).

◀──

endogenous LC339 fluorescence micrograph (right) of unfixed slices from the injected somatosensory cortex taken 1, 4, and 7 d after Ziapin2 administration. The injection site and the diffusion of the compound are visible. Scale bar: 150 μm. (C), Recordings of LFPs evoked in the somatosensory cortex by 20 and 200 ms light stimulation (43 mW/mm^2) in mice injected with either DMSO (black trace) or Ziapin2 (red trace) 1 day before. The cyan-shaded areas represent the light stimulation. Potentials were considered significant above twofold the standard deviation range (broken horizontal lines). (D), Dose-response analysis of LFP responses in DMSO (black)- or Ziapin2 (red)- injected mice for power and duration of the light stimulus (20 and 200 ms; open and closed symbols, respectively). The peak amplitude of light-evoked LFPs has been normalized by the averaged noise amplitude calculated from the nonresponding channels over the same epoch. Photostimulation at increasing power from 4 to 116 mW/mm^2 triggered significant responses in Ziapin2-injected animals that were already significant at 25 mW/mm^2 (200 ms stimulus). No significant responses have been recorded in DMSO-treated animals. (E and F), Time course of the normalized LFP amplitude recorded in the somatosensory cortex 1, 4, and 7 d after intracortical injection of either DMSO (black) or Ziapin2 (red). LFPs have been evoked by 20 (open symbols) and 200 (closed symbols) ms light stimuli at 43 (E) and 116 (F) mW/mm^2. 20 ms, $^{xx}P < .01$, $^{xxx}P < .001$; 200 ms, $^{xx}P < .01$; $^{xxx}P < .001$; repeated measure ANOVA/Tuckey's tests versus DMSO-injected control ($n = 3$ mice for each experimental group) [17].

4.4 Conclusion

This chapter discussed the general concepts of extracellular recordings, spike sorting, and in vivo studies for both photocapacitive and photofaradaic stimulation of neurons with two different technologies of silicon-based photodiodes and polymeric photoswitch particles have been explained. While photofaradaic stimulation provides extracellular potential change through a closed-loop current across the cell membrane's resistance, the photocapacitive stimulation has enabled the change in the cell membrane capacitance with the thickness modulation of the plasma membrane by inserted photoswitch particles inside the plasma membrane. Both technologies have shown reliable stimulation in in vivo condition over long time.

References

[1] Rey HG, Pedreira C, Quiroga RQ. Past, present and future of spike sorting techniques. Brain Research Bulletin 2015;119:106—17.

[2] Gold C, et al. On the origin of the extracellular action potential waveform: a modeling study. Journal of Neurophysiology 2006;95(5):3113—28.

[3] Quiroga RQ. Spike sorting. Scholarpedia 2007;2(12):3583.

[4] Quiroga RQ. Spike sorting. Current Biology 2012;22(2):R45—6.

[5] Homer ML, et al. Sensors and decoding for intracortical brain computer interfaces. Annual Review of Biomedical Engineering 2013;15:383—405.

[6] Quiroga RQ, Nadasdy Z, Ben-Shaul Y. Unsupervised spike detection and sorting with wavelets and superparamagnetic clustering. Neural Computation 2004;16(8):1661—87.

[7] Abeles M, Goldstein MH. Multispike train analysis. Proceedings of the IEEE 1977; 65(5):762—73.

[8] Harris KD, et al. Accuracy of tetrode spike separation as determined by simultaneous intracellular and extracellular measurements. Journal of Neurophysiology 2000;84(1): 401—14.

[9] Shoham S, Fellows MR, Normann RA. Robust, automatic spike sorting using mixtures of multivariate t-distributions. Journal of Neuroscience Methods 2003;127(2):111—22.

[10] Van Drongelen W. Signal processing for neuroscientists. Academic press; 2018.

[11] Oweiss K, Aghagolzadeh M. Detection and classification of extracellular action potential recordings. In: Statistical signal processing for neuroscience and neurotechnology. Elsevier; 2010. p. 15—74.

[12] Dempster AP, Laird NM, Rubin DB. Maximum likelihood from incomplete data via the EM algorithm. Journal of the Royal Statistical Society: Series B (Methodological) 1977;39(1):1—22.

[13] Pal NR, Bezdek JC. On cluster validity for the fuzzy c-means model. IEEE Transactions on Fuzzy Systems 1995;3(3):370—9.

[14] Gray CM, et al. Tetrodes markedly improve the reliability and yield of multiple single-unit isolation from multi-unit recordings in cat striate cortex. Journal of Neuroscience Methods 1995;63(1—2):43—54.

[15] Pedreira C, et al. How many neurons can we see with current spike sorting algorithms? Journal of Neuroscience Methods 2012;211(1):58−65.

[16] Lorach H, et al. Photovoltaic restoration of sight with high visual acuity. Nature Medicine 2015;21(5):476.

[17] DiFrancesco ML, et al. Neuronal firing modulation by a membrane-targeted photoswitch. Nature Nanotechnology 2020:1−11.

Calcium imaging and optical electrophysiology

5.1 Introduction

Recent advances in developing intracellular calcium probes have led to optical mapping as an efficient tool in neuroscience and cardiac research [1,2]. As briefly explained in Chapter 1, generated intracellular calcium waves could program different cell functions such as cell excitation, cell cycle, proliferation, differentiation, and cell death in every cell type in biological organisms [3–5]. In the heart, calcium signaling is used for heart muscle cell contraction [6], and in the nervous system, because of the size and complexity of neural morphology and circuits, calcium signaling has a high degree of versatility. For instance, calcium triggers release of neurotransmitters in presynaptic terminals. Action potentials received at the presynaptic terminal depolarize the plasma membrane, which leads to opening of Ca^{2+} channels and Ca^{2+} flows that stimulate the exocytosis of synaptic vesicles to release their neurotransmitters into the synaptic cleft [7]. In the postsynaptic terminal, transient calcium signals in dendritic spines have been shown to enable induction of activity-dependent synaptic plasticity [8]. Through electrical activation of neurons, calcium ions as the principal second messengers can regulate gene transcription [9]. Moreover, the timescale of intracellular calcium signals varies over a wide range of time, from the microsecond scale (neurotransmitter release to gene transcription) up to minutes and hours [10]. Therefore, different parameters of calcium signals such as the amplitude, time duration, and, most importantly, the location site of their activity in the cell are critical for the function of intracellular calcium signals. In this regard, calcium imaging is important for direct investigation of the diverse neuronal calcium functions, thanks to technological developments in the visualization and quantitative estimation of intracellular calcium signals.

The historical developments in calcium imaging start from both development in fluorescent imaging instruments and fluorescent probes. Ca^{2+}-regulated photoproteins obtained from certain bioluminescent coelenterates such as the *Aequorea* were the first group of calcium indicators, which enabled monitoring of the dynamics of cellular calcium signaling [11,12]. The second group of calcium indicators developed inspired by bioluminescent photoproteins was a synthetic compound (arsenazo III) consisting of an absorbance dye that its light absorption spectrum is a function of calcium bounding [13]. Then more sensitive and versatile fluorescent calcium indicators and buffers were developed, based on hybridization of highly calcium-selective chelators like EGTA or BAPTA with a fluorescent chromophore, by Roger Tsien and colleagues [14]. The first generation of these indicators consists of quin-2, Fura-2, indo-1, and fluo-3.

These fluorescent probes have a different light excitation spectrum and calcium affinity. For instance, the excitation wavelength of fura-2 is at 350 and/or 380 nm and it shows calcium-dependent fluorescence changes in comparison with quin-2. In addition, Fura-2 allows more quantitative calcium measurements with the rationing of signals obtained from two excitation wavelengths [14]. In recent years, advanced indicators with wider excitation spectra and higher calcium affinities have been developed. Oregon Green BAPTA and fluo-4 dye families are among them [15]. These indicators are very popular in neuroscience because the dye loading is relatively easy, and the signal-to-noise ratio is large. Finally, the last class of calcium indicators is based on the introduction of protein-based genetically encoded calcium indicators (GECIs) [16]. GECIs are cell-specific probes that can be used with fluorescent voltage probes at the same time. Although these probes showed slow time response and low signal-to-noise ratios which limited their applications in neuroscience, there have been progresses in solving their limitations [17,18].

On the other hand, optogenetics enables us to control neural activities by depolarizing and hyperpolarizing of the cell membrane through genetically induced photosensitive ion channels and pumps (opsins). This technology has revolutionized the field of neuroscience. Thanks to its high temporal resolution and remote control of neural activities, it enabled researchers to investigate the role of single neurons and neural circuits in neural functions such as vision, learning and memory, and anxiety. In addition, optogenetics has been employed for treatment of retinal prothesis, chronic pain, and cardiac arrhythmia treatment [1]. Moreover, utilizing optogenetics and optical fluorescent monitoring of voltage or intracellular calcium enables "all-optical" electrophysiology, which allows precise activation of neural tissues with high spatiotemporal resolution imaging of action potential and calcium transient morphology and conduction patterns. In this chapter, basic information about optogenetics, florescent probes, and their compatibility with opsins based on their excitation spectrums and their usage for all optical electrophysiology have been provided. In addition, since two-photon microscopy by Winfried Denk and colleagues [19] provided significant advances in high resolution and 3D in vivo calcium imaging in the nervous system [20], basic information about two-photon microscopy has been included in this chapter.

Finally, after providing fundamental information about fluorescent probes, neuronal calcium imaging, optogenetics, and optical electrophysiology, brief guidance in selecting the appropriate calcium indicator and voltage sensors, different dye-loading approaches, and proper device configurations has been provided.

5.2 **Neuronal calcium imaging**

Calcium is an essential intracellular signaling molecule and a second messenger that rapidly affects the kinetics of many cellular processes. Most neurons have an intracellular calcium concentration of about 50–100 nM. This amount can increase transiently during electrical activation by 100 times [3]. There are many different sources of calcium (that regulate calcium influxes and effluxes), including voltage-gated calcium channels, N-methyl-D-aspartate receptors, calcium-permeable AMPA receptors, metabotropic glutamate receptors, and calcium release from internal stores, which are the most important sources of neuronal calcium signaling (as shown in Fig. 5.1). Calcium influxes and effluxes, plus the calcium signaling within internal stores, determine cytosolic calcium concentration. In addition to the aforementioned sources of calcium, cytosolic Ca^{2+}-binding proteins, such

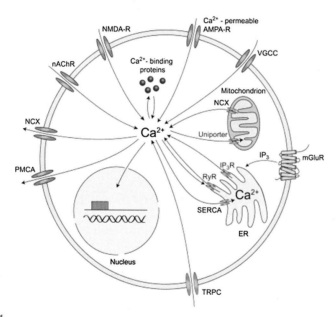

FIGURE 5.1

Different neuronal calcium sources contributing to calcium signaling. Calcium-permeable α-amino-3-hydroxy-5-methyl-4-isoxazolepropionic acid (AMPA) and N-methyl-D-aspartate (NMDA) glutamate-type receptors, voltage-gated calcium channels (VGCC), nicotinic acetylcholine receptors (nAChR), and transient receptor potential type C (TRPC) channels are sources of calcium influx. Inositol trisphosphate receptors (IP$_3$R) and ryanodine receptors (RyR) are mediating calcium release from internal stores. Generation of the inositol trisphosphate can be through metabotropic glutamate receptors (mGluR). Plasma membrane calcium ATPase (PMCA), the sodium-calcium exchanger (NCX), and the sarco-/endoplasmic reticulum calcium ATPase (SERCA) are calcium efflux mediators. Mitochondrion as a Ca^{2+} buffer is also important for neuronal calcium homeostasis [29].

as parvalbumin, calbindin-D28k, or calretinin, act as calcium buffers and modulate temporary intracellular Ca^{2+} signals [21]. The extracellular sources of calcium include voltage-gated calcium channels, ionotropic glutamate receptors, nicotinic acetylcholine receptors (nAChR), and transient receptor potential type C (TRPC) channels [22−24]. To remove calcium ions from the cytosol, plasma membrane Ca^{2+} ATPase (PMCA) as a transport protein in the plasma membrane and the NCX are served. *Ca^{2+} release* from intracellular *stores* [10] is mediated by two major *calcium (Ca^{2+}) release* channels of inositol trisphosphate receptors and ryanodine receptors [25]. For instance, activation of metabotropic glutamate receptors leads to generation of inositol trisphosphate [26]. The endoplasmic reticulum (ER) plays an important role in calcium ion storage, and protein sorting and processing; and the sarco/endoplasmic reticulum calcium ATPase (SERCA) is served to control the high calcium level in the ER. Moreover, mitochondria also function as an important Ca^{2+} buffer, also important for neuronal calcium homeostasis. *Calcium overloads* in the *ER* and *mitochondria*. Calcium uptake by mitochondria occurs during cytosolic calcium elevations, and then it is released back to the cytosol slowly through sodium-calcium exchange [27]. It is important to note that calcium overload in mitochondria, ER, and other types of calcium dysregulation leads to different cell trigger apoptosis and deaths [28].

5.2.1 Voltage-gated calcium channels

Voltage-gated calcium (Ca^{2+}) channels act as key transducers to induce intracellular Ca^{2+} transients (which trigger different physiological events based on the input membrane potential). Multiple types of Ca^{2+} channels have been discovered based on physiological and pharmacological studies. VGCCs include 10 members in mammals, and based on their threshold voltage of activation, they can be categorized into two groups of high- and low-voltage-activated channels [30]. The subfamily of *CaV* channels can be classified as L-, P/Q-, N-, and R-type calcium channels. As discussed in Chapter 1, *CaV α1* is the main subunit that has three major subunits: Cav1, Cav2, and Cav3 [31]. L-type channels (termed Cav1.2, Cav1.3, and Cav1.4) and a skeletal muscle-specific isoform, Cav1.1, are from the Cav1 channel family [32]. The *CaV1 subfamily* (L-type) is expressed in many electrically excitable tissues and mainly contributes to excitation-contraction coupling and is responsible for skeletal muscle contraction. L-type calcium channels in muscles are characterized to be activated at large depolarization (high voltage), having large single-channel conductance and slow inactivation [33,34]. L-type VGCCs also play an important role in neurotransmission of visual signals from *photoreceptors* to *second-order retinal. L-type* channels are *predominantly expressed* in proximal dendrites, where they are involved in dendritic Ca^{2+} signaling resulting from back-propagation of action potentials [35,36] and by synaptically mediated depolarization of dendritic spines [37,38].

Furthermore, other types of voltage-gated channels also can modulate intracellular calcium. One example is the signaling links between Ca^{2+} and Na^+ through sodium calcium exchanger (NCX) pumps and second messengers [39]. As the sodium channels are activated by depolarization signals, the intracellular calcium signal could be altered from the other voltage-gated channels.

5.2.2 N-methyl-ᴅ-Aspartate receptors

Calcium influx response of NMDA receptor channels to neuronal activities only happens when glutamate release and a postsynaptic depolarization occurs at the same time. These calcium influx responses are short-lived signals. The calcium influxes at the postsynaptic terminal of various neuronal cell types, such as pyramidal neurons of the hippocampus [37,40−42] and cortex [43,44], regulate synaptic transmission, plasticity, and cognition in the brain. In general, NMDA receptor channels are nonselective cation channels. They are permeable to sodium, potassium, and calcium ions. Their permeability to calcium ions is about 6%−12% in terms of the fraction to the total cation current [45−48]. However, based on the phosphorylation status of the receptor and the membrane potential of the neuron, they mediate calcium influxes temporarily. Increase of phosphorylation enhances the permeability, while dephosphorylation decreases it [49,50]. Membrane potential as an important factor modulates the efficacy of the voltage-dependent block of NMDA receptors by magnesium [51,52]. Therefore, neural depolarization increases NMDA receptor-dependent ionic current. Another variable that increases the diversity of calcium influx responses is their different subunit compositions [53].

5.2.3 Calcium-permeable α-amino-3-hydroxy-5-methyl-4-isoxazolepropionic acid receptors

GluA2 subunit lacking AMPA receptors is another category of ionotropic glutamate receptor that is permeable to calcium ions. One interesting application of these receptors is their role in synaptic calcium signaling and initiation of synaptic plasticity [54]. In addition, they are expressed in spine-lacking neurons for which physically there is no barrier for generating calcium signal domains in response to single-synapse activation. Their diversity in their subunit composition in a synapse-specific manner is wide [55], and this feature enables different types of responses to distinct synaptic inputs. In addition, permeability of AMPA receptors to calcium varies within a given neuron. This leads to synaptic plasticity mechanisms in aspiny neurons.

5.2.4 Metabotropic glutamate receptors

Metabotropic glutamate receptors (*mGluRs*) are a type of G protein-coupled receptors (GPCRs) connected to various second messenger systems via G proteins [56]. There are different types of I, II, and III mGluRs, and they are expressed in a cell type-specific manner to contribute to diverse physiological events [57]. For example, the activation of mGluR1 subtype can lead to an increase in intracellular calcium in cerebellar Purkinje neurons [58]. After that, glutamate binds to mGluR1, and release of phosphorylate C leads to the generation of IP3, which binds to ER. Therefore, the *release* of intracellular Ca^{2+} stores from the *ER increases intracellular calcium.*

5.2.5 Calcium release and uptake from internal stores

Calcium release and uptake is responsible for the generation of calcium waves and signals. In fact, periodic release of Ca^+ from internal stores leads to alternation of cytoplasmic Ca^+. Two different mechanisms cause calcium release from intracellular stores: (1) Ca^+-induced Ca^+ release (CICR) and (2) IP3Rs-induced Ca^+ release.

Calcium uptake and CICR: CICR is a process in which promotion of calcium leads to release of calcium from intracellular calcium stores. Both calcium uptake and CICR release flux can be described by Michaelis and Menten functions with fixed time constant [59]:

$$\frac{dJ_k\left(\left[Ca^{2+}\right], t\right)}{dt} = \frac{J_{k,\infty}\left(\left[Ca^{2+}\right] - J_k\left[Ca^{2+}\right], t\right)}{\tau_k} \tag{5.1}$$

where $k = U$, CICR.

$$JU_\infty = V_U \frac{\left[Ca^{2+}\right]_i^2}{K_U^2 + \left[Ca^{2+}\right]_i} \tag{5.2}$$

$$J_{CICR_\infty} = V_{CICR} \frac{\left[Ca^{2+}\right]_i}{K_{CICR} + \left[Ca^{2+}\right]_i} \left(\left[Ca^{2+}\right]_s + \left[Ca^{2+}\right]_i\right) \tag{5.3}$$

where V_U and V_{CICR} indicate the maximum rates of uptake and release, respectively.

IP$_3$-induced calcium release: IP$_3$-induced calcium release has been molded with an activation of gate m and an inactivation of gate h, which can be used in neuronal models [59]:

$$J_{IP_3} = V_{IP_3} m^3 h^3 \left(\left[Ca^{2+}\right]_s + \left[Ca^{2+}\right]_i\right) \tag{5.4}$$

$$m_\infty = \frac{[IP3]_i}{[IP3]_i + dIP_3} \frac{\left[Ca^{2+}\right]_i}{\left[Ca^{2+}\right]_i + d_{act}} \tag{5.5}$$

$$h_\infty = \frac{Q[IP_3]}{Q([IP_3]) + \left[Ca^{2+}\right]_i} \tag{5.6}$$

5.3 Fluorescent imaging

Fluorescence is a light emission from fluorescent dyes under an incoming light absorption. The process of fluorescence emission includes (1) absorption of incoming light (photons) by a fluorescence molecule that leads to electrical charge separation and (2) transition of electrons to an excited state (higher energy levels). Afterward, in the absence of incoming light, separated charges in the fluorescence dye start to recombine again. The fluorescent emission occurs after the recombination of the electrical charges. They emit fluorescent light in the form of decay from this excited state.

5.3.1 Excitation

As the Jablonski diagram in Fig. 5.2 shows, incoming photons have the energy of hv coming from an external source such as a lamp or a laser. Then a fluorophore absorbs this photon, and it creates an excited electronic state (S1). This process is physical, and in the absence of light, the fluorescence molecule goes to the ground state.

5.3.2 Excited state lifetime

The fluorescence molecule stays in an excited state for a very short half-life, usually on the order of a few nanoseconds. After this brief period when the light pulse is over, the excited molecules relax toward the lowest vibrational energy level (as shown in Fig. 5.2). This relaxation or coming to the lower energy state results in two products. The first product is the emission of fluorescent and the second is the molecule vibration or heat. The energy lost in this relaxation is dissipated as heat. The energy of the emitted photon (hv) depends on the difference in the energy levels between the two states, and that energy difference defines the wavelength of the emitted light (λ).

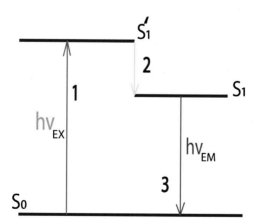

FIGURE 5.2

Jablonski diagram illustrating the processes involved in creating an excited electronic singlet state by optical absorption and subsequent emission of fluorescence: excitation, vibrational relaxation, and emission.

5.4 Calcium indicators

As discussed in the introduction, three different types of bioluminescent, synthetic, and genetically encoded sensors are utilized in the monitoring of intracellular calcium. Fig. 5.3A describes the elements and function of the bioluminescent calcium indicator *aequorin*, derived from jellyfish *Aequorea victoria* [12]. The bioluminescence is a result of both green fluorescent protein (GFP) and a chemi-luminescent protein called aequorin. Aequorin is used as a calcium probe and GFP as a fluorescent marker protein. Aequorin consists of a calcium-binding *apoprotein* (*apoaequorin*) that is covalently bound to a lipophilic cofactor (coelenterazine) [60]. When calcium ions bind to the calcium sites (shown in Fig. 5.3A), conformational change in the protein causes the oxidation of coelenterazine to coelenteramide. Consequently, due to the decay of coelenteramide from the excited to the ground state, a photon is emitted at about 470 nm wavelength [60]. The cytosolic calcium concentration defines the rate of reaction [61]. One of the important advantages of aequorin is its high signal-to-noise ratio and wide dynamic range of cytosolic calcium measurement from 10^{-7} to 10^{-3} M [62]. Another advantage of using biolu-minescent dye is that for recordings of calcium, there is no need of external illumi-nation. This avoids issues such as phototoxicity, photobleaching, autofluorescence, and undesirable stimulation of photobiological processes [63]. However, it has a slow time response as the recharging process with the coelenterazine takes time [64].

5.4.1 Synthetic calcium indicators

The second group of calcium indicators are synthetic or chemical sensors. Fura-2 (shown in Fig. 5.3B) is an example of the fluorescent chemical (or synthetic) calcium indicators [65]. Fura-2 includes two elements of calcium *chelating agent* and fluorophore. Its excitation spectrum is in the range of ultraviolet wavelength (e.g., 350/380 nm), and its emission peak is in the range of green wavelength between 505 and 520 nm [66]. Upon binding of calcium ions, intramolecular confor-mational changes cause fluorescent emission. Fura-2 has dual wavelength excitation, which provides us with the quantitative measurement of the calcium concentration in a neuron independent of the intracellular dye concentration [67].

5.4.2 Genetically encoded sensors

GECIs are used for specific-cell expression [68,69], and they include two types of Förster resonance energy transfer (FRET) (Fig. 5.3C) and the single-fluorophore ones (Fig. 5.3D). *Yellow Cameleon (YC) 3.60* is an FRET-based GECI [70]. The basic working mechanism of FRET is based on a nonradiative energy transfer (electron transfer) between an excited donor fluorophore and an acceptor fluorophore [71]. These two fluorophores should be closer than 10 nm for highly efficient FRET. It includes the enhanced cyan fluorescent protein (ECFP) as donor and the circularly permuted Venus protein as acceptor. The sequence of the

(A) Bioluminescent protein

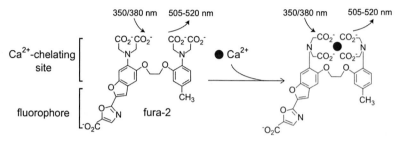

(B) Chemical calcium indicator

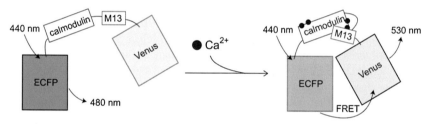

(C) FRET-based GECI

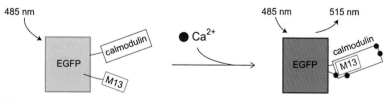

(D) Single-fluorophore GECI

FIGURE 5.3

Calcium fluorescent probes and their working mechanisms. (A) Bioluminescent protein. Binding of calcium ions to aequorin leads to oxidation of the prosthetic group coelenterazine (C, left side) to coelenteramide (C, right side). Coelenteramide relaxes to the ground state while emitting a photon of 470 nm. (B) Chemical calcium indicator. Fura-2 is excitable by ultraviolet light (e.g., 350/380 nm) and its emission peak is between 505 and 520 nm. The binding of calcium ions by Fura-2 leads to changes in the emitted fluorescence. (C) FRET-based genetically encoded calcium indicator (GECI). After binding of calcium ions to Yellow Cameleon 3.60, the two fluorescent proteins, ECFP

calcium-binding protein calmodulin and the calmodulin-binding peptide M13 forms the linker between the two fluorophores [70]. When there are no calcium ions, the emission is close to the blue ECFP fluorescence (480 nm), while upon calcium binding, the spatial distance between the two fluorescent proteins decreases due to the intramolecular conformational changes. Therefore, FRET leads to the Venus protein excitation and consequently causes an emission of about 530 nm. A ratio between the Venus and the ECFP fluorescence is used to measure the calcium signal. The single-fluorophore sensors, GCaMP, have attracted researchers for in vivo experiments [72−75]. Fig. 5.3D shows the structure of GCaMP, which is made by fusion of calmodulin, and the M13 domain consists of a circularly permuted enhanced green fluorescent protein (EGFP) [76]. Upon calcium binding, calmodulin-M13 interactions elicit conformational changes in the fluorophore environment that lead to an increase in the emitted fluorescence [76,77].

Table 5.1 lists the most widely used calcium indicators with examples of their applications and references. One of the most important properties of calcium indicators is their affinities for calcium [15,78]. Calcium affinity is calculated based on the dissociation constant (K_d) describing the probability of separation after Ca^{2+} *ions bind* to this *complex*. Among the calcium indicators, fluo-5N are low-affinity and Oregon Green BAPTA-1 are high-affinity calcium indicators. The calcium affinity or measured K_d value is affected by many parameters such as pH, temperature, and the presence of magnesium [79]. Therefore, it is not the same for in vitro and in vivo condition. It is important to select the appropriate indicator in the appropriate concentration for highest accuracy of the analysis. The design of the calcium measurements should be based on the scientific goals for the type of cells. For instance, as low-affinity fluorescent signals add little buffer capacity to the cell, they provide us with more accurate and faster time response in the free cytosolic calcium concentration [80]. However, the sensitivity of low-affinity calcium indicators is lower than that of the high-affinity one. Sensitivity becomes very critical under noisy conditions like in vivo experiments and when imaging small structures such as dendritic spines. Therefore, there is always a trade-off between sensitivity and response time. Moreover, in calcium imaging experiments, the calcium indicators act as an exogenous calcium buffer and so contribute to the total amount of cellular calcium buffer molecules. Therefore, adding a calcium indicator will change the intracellular calcium dynamics [81]. The single-compartment model considers the endogenous calcium-binding proteins and the exogenous calcium indicator, which enable us to find the unperturbed portion of the calcium dynamic within the cytosol [82,83].

(donor) and Venus (acceptor), approach. This enables Förster resonance energy transfer (FRET), and thus, the blue fluorescence of 480 nm decreases, whereas the fluorescence of 530 nm increases. (D) Single-fluorophore GECI. After binding of calcium to GCaMP, conformational intramolecular changes lead to an increase in the emitted fluorescence of 515 nm [29].

Table 5.1 Frequently used fluorometric calcium indicators [29].

Name	K_d (nM)	Examples of in vivo applications	Representative references
Chemical calcium indicators			
Oregon Green BAPTA-1	170	Mouse neocortex, mouse hippocampus, mouse olfactory bulb, rat neocortex, rat cerebellum, ferret neocortex, cat neocortex, zebrafish	Refs. [73,84–90]
Calcium Green-1	190	Mouse neocortex, mouse olfactory bulb, honeybee, turtle, zebrafish, rat neocortex	Refs. [91–96]
Fura-2	140	Mouse neocortex	Ref. [97]
Indo-1	230	Mouse neocortex	Ref. [98]
Fluo-4	345	Mouse neocortex, *Xenopus* larvae	Refs. [99,100]
Rhod-2	570	Mouse neocortex, zebrafish	Refs. [101,102]
X-rhod-1	700	Mouse neocortex	Ref. [103]
Genetically encoded calcium indicators			
Camgaroo-1		*Drosophila*	Ref. [104]
Camgaroo-2		*Drosophila*, mouse olfactory bulb	Refs. [104,105]
Inverse-Pericam	200	Zebrafish, mouse olfactory bulb	Refs. [105,106]
GCaMP2	840	Mouse olfactory bulb, mouse cerebellum	Refs. [74,107]
GCaMP3	660	Mouse neocortex, mouse hippocampus, *Drosophila*, *Caenorhabditis elegans*	Refs. [73,77,108]
Yellow Cameleon 3.6	250	Mouse neocortex	Ref. [109]
Yellow Cameleon-nano	15 –50	Zebrafish	Ref. [110]
D3cpV	600	Mouse neocortex	Ref. [111]
TN-XL	2200	*Drosophila*, macaque	Refs. [112,113]
TN-L15	710	Mouse neocortex	Ref. [114]
TN-XXL	800	*Drosophila*, mouse neocortex	Ref. [115]

K_d dissociation constant in nM. K_d values taken from the Molecular Probes Handbook (chemical calcium indicators) [70,77,110,115–117].

5.5 Dye-loading approaches

Depending on the type of calcium indicator, the preparation, loading of the dye into neurons, and the measurement could be different. Herein, a simple protocol of calcium dye loading for cortical neurons based on Fura-2 (Fura-2 AM) has been reviewed:

Acetoxy-methyl-ester Fura-2 (Fura-2 AM) is loaded to the cell media to diffuse across the cell membrane. After that, it is deesterified by cellular esterases to yield Fura-2 free acid. As the necessary dye concentration and incubation times for Fura-2 loading vary widely across cell types, it is suggested to prepare several loading solutions with different concentrations of Fura-2 ranging from 1 to 4 μM and incubate the cells for a variety of times from 15 min to 2 h and test the loading at room temperature and at 37°. The following steps are the protocol of calcium dye loading to cortical neurons [118]:

1. **Preparing 1 mM stock solution for the dye**: By adding 50 μL of DMSO to a 50 μg vial. The DMSO should be taken by needle under nitrogen to prevent hydration of the DMSO. The prepared Fura-2 AM solution should be kept t in a dark dry place, and it will be stable at RT for 24 h and several months at −20.
2. **Preparing loading solution**: Adding 2 μL of Fura-2 AM stock solution to the 2 mL of culture media for a 1 μM Fura-2 AM final solution. Then vortex the tube containing the loading dye solution vigorously for 1 min.
3. **Transferring the loading solution to tissue culture dish**: The coverslip with the cells is transferred into the dish. In the case of cultured cells, the cell medium is removed and then the loading solution is replaced.
4. **Incubating the cells in loading solution**: The neurons are incubated at 37° for 30 min in a dark incubator.
5. **Removing loading solution and replacing with cell culture medium**: Fresh complete tissue culture media is prepared in a dish without Fura-2 AM. The coverslip is removed from the loading solution and placed in the new dish. A washing step with calcium-free *phosphate-buffered saline* (PBS) solution before transferring the fresh cultured media is recommended.

Single-cell calcium imaging enables high resolution and real-time analyses of *individual cells* and even their subcellular compartments such as dendrites and spines (for specific examples and application protocols [119]). In addition, calcium imaging is also widely used for studying neural activities in a network of interconnected neurons, for examples, for the analyses of the circuitry of the cortex [120−122], the hippocampus [123], and the retina [124].

Fig. 5.4 illustrates the three most widely used ways of single-cell dye loading. In old imaging experiments, neurons were stained with chemical calcium dyes through sharp microelectrodes [125] and in vivo (Fig. 5.4A, left panel) [96]. Recently, whole-cell patch-clamp micropipettes are used to deliver dye to a single-cell for many different applications (Fig. 5.4A, middle panel) [126,127]. The third approach of single-cell staining is through single-cell electroporation, which is a nonviral method for gene delivery [82,128−130]. In the electroporation technique, applying a few electrical current pulses with appropriate polarity enables dye delivery to the cell (Fig. 5.4A, right panel). The mechanism of dye transfer into the cell is based on two mechanisms: (1) electrical current disrupts the cell membrane and creates pours for diffusing the dyes inside the cell, and (2) the electrical force pushes the charged indicator molecules out of the pipette. This approach also can be used for DNA encoding for GECIs.

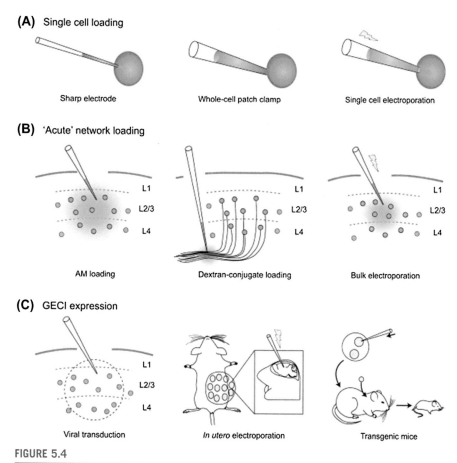

FIGURE 5.4

Dye-Loading techniques. (A) Single-cell loading by sharp electrode impalement (left panel), whole-cell patch-clamp configuration (middle panel), and single-cell electroporation (right panel). These techniques can be used for loading of chemical and genetically encoded calcium indicators (GECIs). (B) "Acute" network loading. A group of neurons are labeled at the same time by acetoxymethyl ester (AM) loading (left panel), via loading with dextran-conjugated dye (middle panel), and by bulk electroporation (left panel). (C) Expression of GECIs by viral transduction (left panel), in utero electroporation (middle panel), and generation of transgenic mouse lines (right panel) [29].

In some studies, calcium imaging is needed for a group of neurons in an intact tissue. Fig. 5.4B (left panel) shows transferring dye through a pressure pulse for targeting bulk dye loading of membrane-permeable acetoxymethyl (AM) ester calcium dyes [65] based on multicell bolus loading [98]. In this simple method, an AM calcium dye (for example, Oregon Green BAPTA-1 AM), already in a dye ejection pipette, is delivered by an air pressure pulse to brain tissue, which results in a stained area with a diameter of 300–500 μm [98,131,132]. In addition to AM calcium dyes, other types of calcium dyes such as dextran-conjugated chemical calcium indicators can also be transferred by using pressure injection to axonal pathways (Fig. 5.4B, middle panel) [133]. This method has already been used for staining neural networks and calcium imaging from axonal terminals in the mouse cerebellum and olfactory bulb [92,94,134] as well as calcium signals in spinal cord neurons [135]. Furthermore, electroporation can be employed (see above) to deliver calcium dyes to local neuronal networks (Fig. 5.4B, right panel) [103]. The same as explained for single-cell staining, a micropipette containing the dye in salt form or as dextran conjugate is inserted into the brain or spinal cord area of interest, and trains of electrical current pulses are applied to transfer the dye to nearby cell bodies. This technique has been successfully used for in vivo studies in mouse neocortex, olfactory bulb, and cerebellum [103,136].

In recent decade, GECIs have been utilized as promising imaging probes in neuroscience for imaging calcium dynamics and neuronal activities in vivo [17]. The most popular approach for expressing GECIs in neurons is viral transduction (Fig. 5.4C, left panel). *Stereotactic injection* is a method to target specific brain areas for staining neurons with GECIs [137]. In this method, genetically engineered of viral vectors such as enti- (LV) [138], adeno-associated [139], herpes simplex [140], and recently ΔG rabies are employed to introduce GECIs into the host cells [141]. Utero electroporation is another method for controlled spatiotemporal gene transfection, and it has been applied for expression of the GECI through DNA plasmids encoding. In this method, in contrast to viral delivery, the labeling is in a relatively more sparse manner (Fig. 5.4C, middle panel) [115]. In utero electroporation, as in single-cell and bulk electroporation techniques, an electrical potential (force) is applied to push negatively charged DNA molecules into the cells [142]. It is important to note that this method does not have a size limitation for the transfected gene of interest, and that it can be used for species where transgenic technology is not easily implemented. Finally, in recent years, there have been some efforts to apply transgenic mice to express GECIs [70,143–145] (Fig. 5.4C, left panel). For instance, transgenic mice expressing the sensitive red GECI jRGECO1a, driven by the *Thy1* promoter, have been used for calcium imaging of functionally diverse ganglion cell populations in the retina [146]. Stable transgenic mouse lines with two different functional calcium indicators, inverse-pericam and camgaroo-2, under the control of the tetracycline-inducible promoter, have been reported for calcium imaging in the mouse olfactory bulb in vivo [105].

At the end, it would be useful to compare chemical calcium indicators and GECIs. Chemical calcium indicators have high signal-to-noise ratio and fast kinetics [80,147]. In addition, dye loading approaches for the chemical calcium indicators both for single cells or small groups of cells are well established, similar for various types of mammalian cells, and relatively easy (Fig. 5.4A and B). One drawback of the chemical indicators is that it is difficult to stain cells in a specific-cell manner—for instance, for a specific type of interneurons. The second serious issue of using chemical indicators is related to the limiting time of recordings, and it makes it difficult to perform long chronic recordings [148]. In this kind of long experiment, GECIs have been shown to be functional in neurons over long periods [77,115,148].

In contrast to chemical indicators, GECIs enables molecularly defined cell types or even subcellular compartment measurements [149−152]. However, they have rather slow kinetics [147]. Expressing GECIs through the aforementioned techniques such as utero electroporation can cause tissue damage and cytotoxicity in the long term [77].

5.6 Analysis of calcium images and videos

In general, recorded videos from fluorescent stained cells are analyzed frame by frame. In this regard, cells should be detected, and the analysis should be applied to all frames of the video to track the fluorescent intensity change coming from cells. The region of interest (ROI) technique has been largely used for extracting cellular signals from Ca^{2+} imaging data. In this method, with drawing a ROI around the cells, the light intensity of the dye in the region could be traced in time. In fact, manual means of identifying cells are not easy to handle for largest calcium imaging datasets. It requires undue human labor. In addition, this type of analysis is prone to cross talk coming from adjacent cells. Nowadays, thanks to automated spike sorting, calcium spikes can be assigned to individual cells, which enables us to understand neural coding. Automated spike sorting for tracing single-cell intracellular calcium can be performed based on some logic. First, during the cellular events such as action potential generation, the intracellular calcium of cells changes in comparison to the background signal. Second, elevated intracellular calcium occurs in a short period of time, which is distinguishably dispersed among background time frames. Considering that each cell occupies only a small area of pixels, intracellular calcium signals are distinguishable in spatial domains. Therefore, intracellular calcium waves are distinguishable by a distribution of amplitude.

The aforementioned logics help us to identify cells' activities individually by extracting data into a summation of statistically independent signals, each with a high resolution. Fig. 5.5 shows a spike-sorting algorithm based on an independent component analysis (ICA). ICA is a smart algorithm for extracting signals coming from several neighboring sources, and it has been used already for analyses of electroencephalography [153], magnetoencephalography, and functional magnetic resonance imaging data. This algorithm includes principle component analysis (PCA). PCA provides us with data visualization to identify patterns and calcium waveforms more easily. It also helps to reduce dimensions to decrease computation and lower error rate.

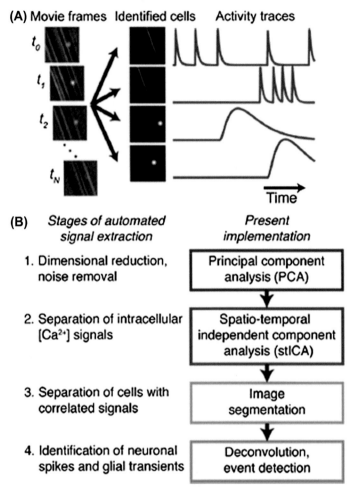

FIGURE 5.5

Analytical protocol for automated cell sorting. (A) Cell sorting includes different stages of extracting cellular signals from imaging data (left) through estimating spatial filters (middle) and activity traces (right) for each cell. The figure shows an example depicting typical fluorescence transients in the cerebellar cortex as observed in optical cross section. Calcium transient signals in Purkinje cell dendrites are coming from across elongated areas seen as stripes in the movie data. Transients in Bergmann glial fibers appear be more *localized* Ca^{2+} sparkles and ellipsoidal waveform. (B) Automated cell sorting consists of four steps, which aimed at addressing specific analysis challenges [154].

5.6.1 Principles for extracting cellular signals

Cell sorting is widely used to extract time traces related to the individual cells in different locations of the image frames by using spatial filters (Fig. 5.5A). The procedure of cell sorting includes the following four stages [154]:

1. Dimensional reduction:
   In PCA, an orthogonal *linear* transformation of the data is identified to find a pattern based on a new set of basis vectors (the principal components that present more variance from the dataset). PCA also reduces redundancy and so discards dimensions that mainly represent noise [155]. Using PCA by itself is not enough for cell sorting because it depends on differences in variance to find data components; however, practically different cells' fluorescence signals tend to have similar amplitude over time. It means that a mix of signals from multiple cells forms each principal component.
2. Spatiotemporal ICA for extraction of Ca^{2+} signals.
3. Image segmentation for separating correlated signals.
4. Temporal deconvolution and spike detection to extract spike times.

5.7 Optogenetics

5.7.1 Tools of optogenetics—opsins

Optogenetics is a technique to optically manipulate (excitation and inhibition) and monitor cellular activities (membrane voltage and membrane conductances) through inducing genetically encoded proteins in the cell membrane. These proteins are light-sensitive ion channels that enable optical control of the ionic currents and membrane permeability to specific ions. Microbial opsins called channelrhodopsins function as light-gated ionic channels and initiated the field of optogenetics in the early to mid-2000s [156,157]. Rhodopsins are GPCRs with *seven transmembrane α-helices* (GPCRs). For example, channelrhodopsin2 (ChR2) is a widely used opsin and provided us with drastic advancement in neuronal modulation through optogenetics. Opening of ChR2 by blue light ($\sim$470 nm) excitation leads ions including Na^+ to enter the cell and leads to cell depolarization. Fig. 5.6 shows a schematic of rhodopsins that can be used as voltage sensors and actuators. Rhodopsins can also be used as genetically encoded voltage indicators (GEVIs). Fig. 5.7 shows the family of GEVIs and their developments over time. This family could be classified based on two groups of GEVIs: voltage-sensor domains and microbial rhodopsin-based GEVIs. In GEVIs indicators, conformational changes directly affect their fluorescence response or through eliciting processes such as FRET. Archaerhodopsin voltage-based sensors (QuasAr1 and QuasAr2) are recently developed GEVI sensors that have shown improved voltage sensitivity and fast response (microsecond) without photocurrent generation.

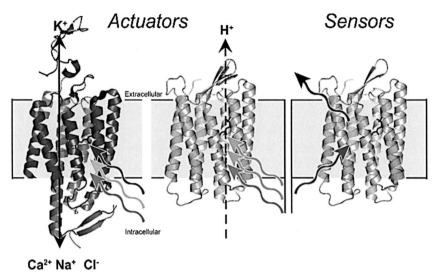

FIGURE 5.6

Rhodopsins are used as actuators and sensors in optogenetics [163]. Actuators facilitate the ionic transport across the cell membrane to activate or inhibit neuronal activity. ChRs transport positively charged ions into the cell, while proton-pumping rhodopsins move protons out of the cell. Ideally, engineered rhodopsin sensors can have fluorescence light emission in the far-red depending on the membrane voltage [163].

Another group of fluorescent voltage probes is synthetic sensors, which cannot specifically stain all the cell subpopulations (similar to what has been explained in calcium indicators), while genetic sensors enable us to specifically target the cells and perform long-term measurements (or other cell types) [158–160]. Di-4-ANEPPS and rh-237 are widely used voltage sensors that show fast time response (femto- to picosecond). However, they typically show a relatively low sensitivity of 2%–10% fluorescence change per 100 mV. Their working mechanism is based on a shift of fluorescent signal as a result of shifts in charge state and hence dipole energy levels (electrochromism) [161,162].

5.7.2 Opsin and sensor compatibility

For a closed-loop system like optical electrophysiology technique, light absorbance and excitation of opsins should not interfere with voltage and Ca^{2+} sensors, as spectral overlap causes unwanted cross talk and perturbation of the cellular membrane. For example, rhod-4AM (Ca^{2+} indicator), which exhibits less spectral overlap with ChR2, is routinely used in all optical setups [165]. In contrast, the excitation spectrum of 4-ANEPPS, which optically measure voltage, overlaps with ChR2, and so excitation of both at the same wavelengths perturbs the membrane potential (Fig. 5.8). Using sensors with red-shifted absorption spectrum is the most common

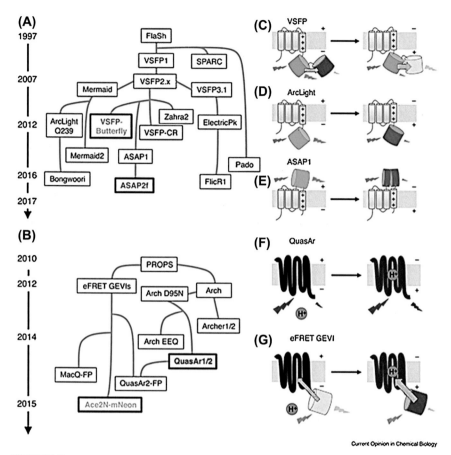

FIGURE 5.7

The family of GEVIs and their evolution over time. (A and B) The major classes of GEVIs with their relative lineage. The highlight in bold shows the GEVIs with greatest sensitivity for use in vivo and are colored with their approximate excitation wavelengths. (A) GEVIs based on voltage sensor domains. (B) GEVIs based on microbial rhodopsins. (C–G) The mechanisms of voltage sensing in the major classes of GEVIs [164].

solution for biocompatibility. In general, employing a red-shift in the excitation spectrum of the sensor facilitates easier signal processing for an all-optical system as a long-pass or band-pass filter can be easily applied to prevent possible cross talk. Table 5.2 reviews different opsins and voltage and calcium sensors that have been already used with optogenetic control and gives examples of illumination and filter setups minimizing actuator-sensor cross talk. As the table illustrates, PGH1, di-4-ANBDQPQ, di-4-ANBDQBS, and rh-1691 have good compatibility in terms of absorption spectrum and were used successfully in all optical setups.

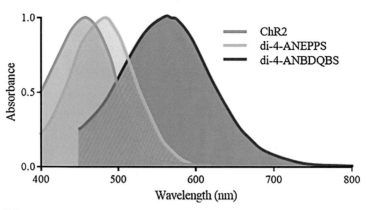

FIGURE 5.8

Fluorescence excitation spectrum of the opsin ChR2-H134 and voltage-sensitive sensors di-4-ANEPPS and di-4-ANBDQBS. In this absorbance spectrum, there is a significant overlap between the excitation spectra of ChR2 and di-4-ANEPPS, which prevents imaging of di-4-ANEPPS without excitation of ChR2. However, the red-shifted spectra of di-4-ANBDQBS provide us with a shifted excitation wavelength preventing simultaneous activation of ChR2 [1].

Moreover, designed "blue-shifted" opsins [166] also facilitates prevention of spectral overlap. However, shorter absorption wavelength limits their usage by tissue damage and penetration depth.

5.7.3 Illumination sources

In optogenetics and fluorescent imaging, using light-emitting diodes (LEDs) is beneficiary as they provide us with narrow wavelength spectra, long operational lifetimes, and low heat emission [182]. Lasers, tungsten-halogen lamps, mercury/xeon, and arc lamps are other illumination sources. Based on the desired wavelength and power of the light source, proper light sources are selected. While for the imaging, spatial and temporal homogeneity is important, optical actuation in optogenetics needs impulse-like signals (temporal inhomogeneity) that are focused on a specific area of the sample (spatial inhomogeneity). Additionality, synchronous patterned light can be generated with liquid crystal and digital micromirror device (DMD) emerged as practical *spatial light modulators* [183]. This type of light modulation has been widely used in cardiac optogenetics [184]. A DMD chip embedded with *several hundred thousand* microscopic *mirrors* to deliver illumination patterns with high spatial resolution has enabled several optogenetic-based discoveries [173].

To focus the light and deliver it over longer distances, fiber optic-coupled LED [185] and laser-based approaches have been employed. This facilitated incorporation of illuminating fibers in clinical tools for precise spatial illumination in vivo [186]. In addition, fully implantable and wireless optogenetic devices embedded with organic light-emitting diodes as biocompatible light sources have also been developed, which are promising for clinical benefits of optogenetics.

Table 5.2 Combinations of opsins and fluorescent sensors used in all-optical setups, with excitation wavelengths if opsin and sensor, and sensor emission spectral characteristics from specific studies [1].

Opsin	Opsin excitation λ (nm)	Sensor	Type	Excitation source	Sensor excitation λ (nm)	Sensor emission λ (nm)	References	Other studies using specified opsin/sensor combination
ChR2 variants: ChR2-H134R, CatCh, CheRiff (depolarizing)	470	Di-4-ANBDQBS	Synthetic	LED	660	LP700	[167]	[168–171]
		Di-4-ANBDQPQ	Synthetic	LED	(655/40) 625	774/140	[172]	[173–175]
		PGH1	Synthetic	LED	(640/40)	LP760	[176]	
		Rh-237	Synthetic	Hg/Xe arc lamp	655	LP650	[177]	[178]
		Rh-421	Synthetic	Halogen lamp	(640/40)	630/69	[174]	
		Rh-1691	Synthetic	LED	(560/55)	–	[179]	
		BeRST1	Synthetic	LED	(635/30)	LP665	[174]	
		Di-4-ANEPPS	Synthetic	Halogen lamp	(525/50)	LP600	[180]	
		QuasAr 1	GEVI	Laser	593.5	LP760	[174]	
		QuasAr2	GEVI	Laser	640	630/69	[159]	[181]
		Arch(D95N)	GEVI	Laser	647		[181]	
Calcium sensors		Rhod-2AM	Synthetic	LED	530	570–625	[167]	[165]
		Rhod-4AM	Synthetic	LED	530	LP565	[178]	
		GCaMP5f	GECI	Laser	488	570–625	[181]	
		GCaMP6f	GECI	Laser	488	LP565	[159]	
Voltage sensors ArchT (hyperpolarizing)	566	QuasAr 1	GEVI	Laser	593.5	LP665	[174]	
eNpHR3.0 (hyperpolarizing)	590	PGH1	Synthetic	LED	655(690/60)	LP760	[176]	

5.7.4 Optical filtering

Optical filtering is advantageous to avoid spectral overlap, which has already been discussed in Section 5.7.2, and so, it is still crucial in effective excitation and imaging of samples, especially in case of multiphoton experiments in which different lights with different wavelengths are used. Single-wavelength filters can be classified as (A) bandpass filters, (B) long- or short-pass filters, and (C) dichroic mirrors/beam splitters. A bandpass filter could be designed to transmit a specific wavelength or wavelengths, which includes a window characterized by a central wavelength and full-width half maximum (FWHM). To narrow excitation wavelength, bandpass filters are used, where light is filtered to narrow spectral bandwidth with a relatively small FWHM filter. This prevents interference of signals among sensors or actuators [187].

5.8 Optical electrophysiology (Optopatch)

Optical electrophysiology uses optogenetics to manipulate cellular functions and fluorescent sensors to track intracellular ionic currents and cell membrane potential. However, accurate changes in membrane potential and measurements of ionic currents are needed, and measurements of current to assay different ion channels. Optical electrophysiology can be a semiquantitative technology in comparison with the patch-clamp. In this regard, using the optical techniques has some challenges: expression levels of optogenetic actuators and sensors are different from cell to cell; channelrhodopsins change conductance rather than membrane potential directly (not a voltage clamp); and fluorescence sensors are able to measure only membrane voltage, not current [188,189]. Therefore, optical electrophysiology cannot be used as an alternative for standard voltage-clamp protocols. However, this technology enables the study of complex interactions underlying neural dynamics, visualizing membrane potentials across a spatially and temporally precise stimulation [190,191].

In this technology, an actuator such as blue light-gated channelrhodopsin CheRiff is paired with a red light-excitation voltage indicator like QuasAr which enables cross talk-free genetically targeted all-optical electrophysiology [11]. Fig. 5.9A shows an engineered HEK293 cell line stably expressing human $Na_V 1.7$ channel and the Optopatch constructs including both QuasAr2-mOrange2 and CheRiff-eGFP, which shows good membrane trafficking (Fig. 5.9B). To assay the $Na_V 1.7$ channel, with activation of the light actuator, the cell is depolarized and so the cell membrane potential changes. Consequently, it causes opening of the sodium channel.

The optical and electrophysiological performance of the indicator, the actuator, and the coexpressed pair (Optopatch) is characterized by patch-clamp measurements. Fig. 5.10 shows the results of the patch-clamp measurements. In voltage-clamp ($V_m = 60$ mV), membrane current is altered by the activation of CheRiff by

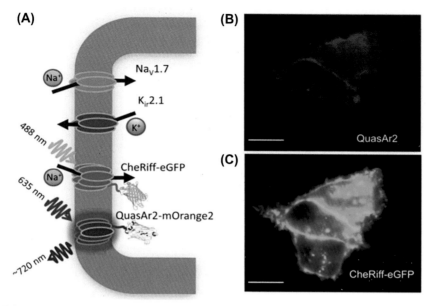

FIGURE 5.9

Optopatch for electrically spiking HEK cells. (A) Expressed HEK cells with $Na_V1.7$-OS (imparts electrical excitability) and $K_{ir}2.1$, which maintains a hyperpolarized resting potential close to the K^+ reversal potential. Upon light excitation, CheRiff as an actuator depolarizes the cells and can stimulate a $Na_V1.7$-mediated action potential. QuasAr2 as the voltage sensor is excited by red light and emits near infrared fluorescence. (B and C) Epifluorescence images taken from QuasAr2 and CheRiff-eGFP under excitation. Scale bar 10 μm [192].

488 nm light with an EPD50 (effective power density for 50% activation) of 20 mW/cm^2, and it reaches a steady-state photocurrent density of 13.0 ± 1.2 pA/pF (mean - ± s.e.m., $n = 5$ cells, Fig. 5.10A and B). The conductance of $Na_V1.7$ channel is investigated under voltage steps from a holding potential of −100 mV. Fig. 5.10B shows robust inward currents with fast activation and inactivation kinetics within 10 ms and a peak current density of 61.4 ± 13.6 pA/pF at 20 mV (mean ± SD, $n = 11$ cells, Fig. 5.10C).

The I-V response of expressed $K_{ir}2.1$ channel also has been shown in Fig. 5.10D. In current-clamp measurements, simultaneous recordings of membrane potential and QuasAr2 fluorescence are studied under optical CheRiff stimulation (as shown in Fig. 5.10E). QuasAr2 fluorescence is activated with continuous red light (640 nm, 400 W/cm^2) and pulses of blue light (500 ms on, 1.5 s off, stepwise increasing intensity from 1.1 to 26.0 mW/cm^2) are used for activation of CheRiff. As Fig. 5.10E shows, fluorescence signals reproduce both the action potential waveforms and the subthreshold depolarizations.

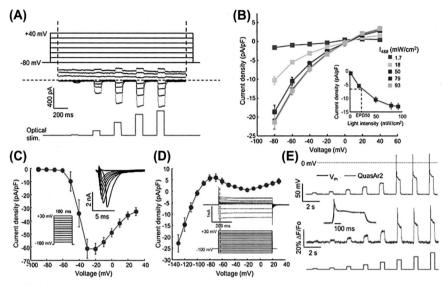

FIGURE 5.10

Biophysical characterization of Na$_V$1.7-OS HEK cells and the performance of Optopatch setup. (A) CheRiff photocurrent induced in a Na$_V$1.7-Optopatch HEK cell. Membrane potential has been lamped at −80 mV and then it is swept for 2 s from −80 to +40 mV with 20 mV steps. Depolarizing signals have been generated with five pulses of blue light, 100 ms duration, with different intensity levels (1.7, 18, 50, 79, 93 mW/cm^2). (B) I-V curve of CheRiff, obtained for different light intensities. (C) Peak Na$_V$1.7 current densities in response to different level of depolarization potentials. Voltage-clamp measurement with −100 mV holding potential and input voltage pulses with 100 ms pulse duration from −90 mV to +30 mV in 10 mV increments. These measurements are for the cell before expression of K$_{ir}$2.1. Inset: recorded currents in the 10 ms interval after each voltage step. (D) I-V curve for K$_{ir}$2.1 expressed in Na$_V$1.7-OS HEK cells. In voltage-clamp protocol, holding potential at −100 mV with pulse duration of 500 ms from −130 mV to +30 mV with 10 mV increments. Inset: representative K$_{ir}$2.1 current recording. *Red line* shows (4 ms after voltage step) when the current has been quantified. (E) Current-clamp recording at the same time with QuasAr2 fluorescence recording from Na$_V$1.7-OS HEK cells. A series of blue laser pulses with 500 ms duration, with increasing intensities (1.1, 2.3, 4.3, 7.0, 11, 15, 20, 26 mW/cm^2), and QuasAr2 fluorescence was monitored with 640 nm excitation, and light intensity of 400 W/cm^2. Inset: shows that the voltage and fluorescence recordings are similar for the most intense blue pulse (26 mW/cm^2) [192].

5.9 Two-photon manipulation and imaging

The benefits of using two-photon microscopy to compare with confocal imaging includes higher resolution since the light is collected near the focal plane where the laser is focused, higher penetration depth because of longer wavelength

($\sim$800 nm) photons, optical sectioning for three-dimensional imaging, and less photodamage as it uses lower illumination intensity [193]. In calcium imaging assisted by two-photon excitation microscopy, Ca^{2+} fluorescence signals could be explored in spines located on dendritic segments deep below the surface about 100–200 μm inside a 300–450 μm thick hippocampal slice. On the other hand, two-photon optogenetics with simultaneous volumetric two-photon calcium imaging enables manipulation and monitoring of neural in vivo and in three dimensions (3D) with cellular resolution [194].

Fig. 5.11 shows the basic principle of one-photon and two-photon excitations. In normal fluorescent microscopy, a photon with an energy of $h\nu$ with a wavelength of λ is absorbed by the fluorophore; on the other hand, in two-photon excitation, two photons with a wavelength twice as long (λ_{2p}) provide the same energy required to achieve an equivalent transition under one-photon excitation. For instance, one photon is absorbed by fluorophore with an absorption at 450 nm in one-photon conventional microscopy to emit the fluorescent emission with an emission photon. In two-photon imaging, this photon could be at approximately 800 nm with half of one photon energy, while the emission fluorescent photon is the same as in the one-photon conventional microscopy.

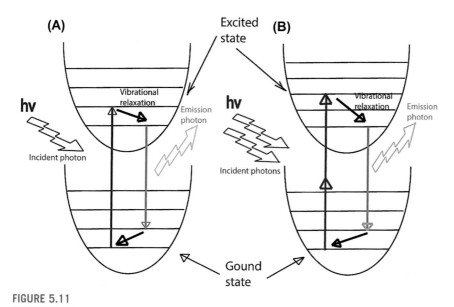

FIGURE 5.11

Jablonski diagrams (A) *for one-photon* excitation and (B) *two-photon* excitation. Excitations occur between the ground state and the vibrational levels of the first electronic excited state.

Moreover, in confocal microscopes, the focal volume of the excitation is limited to the pinhole of the objective, while in two-photon excitation, all the emitted light is collected regardless of the path and assigned to focus without using a pinhole. Fig. 5.12 illustrates the setup and principle of two-photon excitation microscopy. Fig. 5.12A shows that the two-photon excitation (*red lines*) absorbed by a fluorophore leads to one-photon emission (*green line*). In two-photon microscopy, excitation photons are focused on a scattered media suffering from diffraction limit; however, localization of the excitation area enables detection of all collected photons, even scattered ones (Fig. 5.12B and C). Fig. 5.12D shows optical setup and the elements used for a two-photon imaging setup.

5.10 Calcium imaging in neuron cell death

Intracellular Ca^{2+} is involved in apoptosis and neuron cell death, and calcium overload is one example that leads to neuron cell death and excitotoxicity. In this regard, calcium imaging is one approach to define proper cell stimulation parameters through measuring intracellular Ca^{2+} concentration in neurons. Fluorescence imaging of cytosolic Ca^{2+} for Fura-2/AM-loaded neurons and the threshold of calcium increase for the cell death at the same time has been used to detect apoptosis with apoptotic markers such as Annexin V.

The following protocol has been used for correlating Ca^{2+} changes with susceptibility to apoptosis in the same neurons [196]:

1. Incubating cells or coverslips containing cultured hippocampal cells in *phosphate buffer saline* (*PBS*) *solution* with dye loading solution (4 µM Fura-2/AM) for 60 min.
2. Adding an apoptosis-inducing agonist such as NMDA to stimulate the cells and recording increase of $[Ca2^+]_{cyt}$ in real time.
3. Recording the images while using a filter for 520 nm emission and 340 and 380 nm excitation light source.
4. Washing cells with warm PBS at 37°C for 10 min to remove the stimulus. Keeping them in the same medium at room temperature for several hours for assessing apoptosis in the same microscopic field.
5. Staining the cell with apoptotic marker of Annexin V (diluted 1:20 in Annexin V binding buffer): carefully emptying the chamber and by adding drops of Annexin V solution to the coverslip and incubating cells for 10 min.
6. Assessing the stain after washing the cells stained with Annexin V with *PBS* by fluorescence in the same microscopic field using a 40× objective.

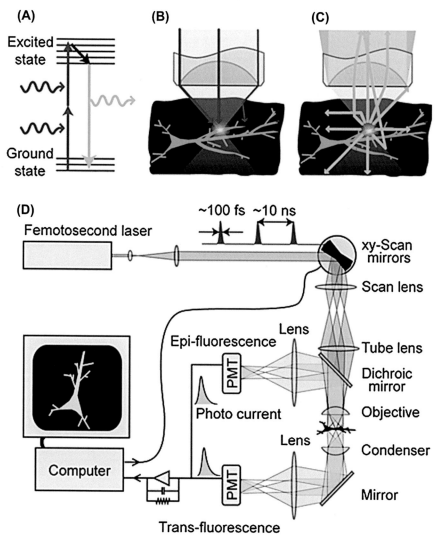

FIGURE 5.12

The principle and setup for two-photon excitation microscopy. (A) Simplified Jablonski diagram for the two-photon excitation process. (B) An excitation beam light is focused through a scattering medium (*black*) to a diffraction-limited spot by an objective to excite green fluorescence in a dendritic branch, but not in a nearby branch. Traveling paths of the two ballistic photons and one scattered photon are shown by red lines. Scattered photons are too weak for off-focus excitation, and as the number of scattered photons increases for deeper thicknesses, the intensity of the beam drops with depth. (C) Contribution emitted photons (*green lines*) in a scattering medium. Even scattered fluorescence photons contribute to the signal if they are collected by the objective. Since in two-photon microscopy the field of view for detection is larger than for excitation (does not depend on the pinhole of the objective), the fluorescence light exiting the objective back aperture will diverge significantly (*green*). (D) The block diagram of a two-photon excitation microscope with epifluorescence and transfluorescence detection [195].

5.11 Conclusion

For monitoring cell stimulation and death, calcium imaging is crucial, and it is also widely used in understanding neural activities. In this regard, a biophysical calcium dynamic model and methodology of calcium imaging have been explained in this chapter, enabling students and scientists in different fields to utilize them in their research. In addition, Optogenetics-based therapy has showed promising results in clinical practice, but still there is a need for technological and methodological improvements. These improvements include delivering light in dipper thickness of tissue, delivering and efficient gene delivery, and remote control of light through implantable devices. However, utilizing optogenetics in all-optical stimulation and imaging systems is already used as a novel and transformative tool for cardiac research and in the study of the physiology and pathophysiology of the heart.

References

[1] O'Shea C, et al. Cardiac optogenetics and optical mapping—overcoming spectral congestion in all-optical cardiac electrophysiology. Frontiers in Physiology 2019;10.

[2] Herron TJ, Lee P, Jalife J. Optical imaging of voltage and calcium in cardiac cells & tissues. Circulation Research 2012;110(4):609−23.

[3] Berridge MJ, Lipp P, Bootman MD. The versatility and universality of calcium signalling. Nature Reviews. Molecular Cell Biology 2000;1(1):11−21.

[4] Lu KP, Means AR. Regulation of the cell cycle by calcium and calmodulin. Endocrine Reviews 1993;14(1):40−58.

[5] Orrenius S, Zhivotovsky B, Nicotera P. Regulation of cell death: the calcium−apoptosis link. Nature Reviews. Molecular Cell Biology 2003;4(7):552−65.

[6] Dulhunty A. Excitation−contraction coupling from the 1950s into the new millennium. Clinical and Experimental Pharmacology and Physiology 2006;33(9): 763−72.

[7] Neher E, Sakaba T. Multiple roles of calcium ions in the regulation of neurotransmitter release. Neuron 2008;59(6):861−72.

[8] Zucker RS. Calcium-and activity-dependent synaptic plasticity. Current Opinion in Neurobiology 1999;9(3):305−13.

[9] Lyons MR, West AE. Mechanisms of specificity in neuronal activity-regulated gene transcription. Progress in Neurobiology 2011;94(3):259−95.

[10] Berridge MJ, Bootman MD, Roderick HL. Calcium signalling: dynamics, homeostasis and remodelling. Nature Reviews. Molecular Cell Biology 2003;4(7):517−29.

[11] Ashley C, Ridgway E. Simultaneous recording of membrane potential, calcium transient and tension in single muscle fibres. Nature 1968;219(5159):1168−9.

[12] Shimomura O, Johnson FH, Saiga Y. Extraction, purification and properties of aequorin, a bioluminescent protein from the luminous hydromedusan, Aequorea. Journal of Cellular and Comparative Physiology 1962;59(3):223−39.

[13] Brown J, et al. Rapid changes in intracellular free calcium concentration. Detection by metallochromic indicator dyes in squid giant axon. Biophysical Journal 1975;15(11): 1155.

[14] Tsien RY. New calcium indicators and buffers with high selectivity against magnesium and protons: design, synthesis, and properties of prototype structures. Biochemistry 1980;19(11):2396−404.

[15] Paredes RM, et al. Chemical calcium indicators. Methods 2008;46(3):143−51.

[16] Miyawaki A, et al. Dynamic and quantitative Ca^{2+} measurements using improved cameleons. Proceedings of the National Academy of Sciences 1999;96(5):2135−40.

[17] Looger LL, Griesbeck O. Genetically encoded neural activity indicators. Current Opinion in Neurobiology 2012;22(1):18−23.

[18] Rochefort NL, Jia H, Konnerth A. Calcium imaging in the living brain: prospects for molecular medicine. Trends in Molecular Medicine 2008;14(9):389−99.

[19] Denk W, Strickler JH, Webb WW. Two-photon laser scanning fluorescence microscopy. Science 1990;248(4951):73−6.

[20] Yuste R, Denk W. Dendritic spines as basic functional units of neuronal integration. Nature 1995;375(6533):682−4.

[21] Schwaller B. Cytosolic Ca^{2+} buffers. Cold Spring Harbor Perspectives in Biology 2010;2:a004051.

[22] Fucile S. Ca^{2+} permeability of nicotinic acetylcholine receptors. Cell Calcium 2004; 35(1):1−8.

[23] Higley MJ, Sabatini BL. Calcium signaling in dendrites and spines: practical and functional considerations. Neuron 2008;59(6):902−13.

[24] Ramsey IS, Delling M, Clapham DE. An introduction to TRP channels. Annual Review of Physiology 2006;68:619−47.

[25] Berridge MJ. Neuronal calcium signaling. Neuron 1998;21(1):13−26.

[26] Niswender CM, Conn PJ. Metabotropic glutamate receptors: physiology, pharmacology, and disease. Annual Review of Pharmacology and Toxicology 2010;50: 295−322.

[27] Duchen MR. Contributions of mitochondria to animal physiology: from homeostatic sensor to calcium signalling and cell death. The Journal of Physiology 1999;516(1): 1−17.

[28] Celsi F, et al. Mitochondria, calcium and cell death: a deadly triad in neurodegeneration. Biochimica et Biophysica Acta (BBA) - Bioenergetics 2009; 1787(5):335−44.

[29] Grienberger C, Konnerth A. Imaging calcium in neurons. Neuron 2012;73(5):862−85.

[30] Catterall WA. Structure and regulation of voltage-gated Ca^{2+} channels. Annual Review of Cell and Developmental Biology 2000;16(1):521−55.

[31] Catterall WA, et al. International Union of Pharmacology. XLVIII. Nomenclature and structure-function relationships of voltage-gated calcium channels. Pharmacological Reviews 2005;57(4):411−25.

[32] Catterall WA. Voltage-gated calcium channels. Cold Spring Harbor Perspectives in Biology 2011;3(8):a003947.

[33] Reuter H. Properties of two inward membrane currents in the heart. Annual Review of Physiology 1979;41(1):413−24.

[34] Tsien R, et al. Multiple types of neuronal calcium channels and their selective modulation. Trends in Neurosciences 1988;11(10):431−8.

[35] Spruston N, et al. Activity-dependent action potential invasion and calcium influx into hippocampal CA1 dendrites. Science 1995;268(5208):297−300.

[36] Waters J, Schaefer A, Sakmann B. Backpropagating action potentials in neurones: measurement, mechanisms and potential functions. Progress in Biophysics and Molecular Biology 2005;87(1):145–70.

[37] Bloodgood BL, Sabatini BL. Nonlinear regulation of unitary synaptic signals by $CaV_{2.3}$ voltage-sensitive calcium channels located in dendritic spines. Neuron 2007; 53(2):249–60.

[38] Reid CA, Fabian-Fine R, Fine A. Postsynaptic calcium transients evoked by activation of individual hippocampal mossy fiber synapses. Journal of Neuroscience 2001;21(7): 2206–14.

[39] Verkhratsky A, et al. Crosslink between calcium and sodium signalling. Experimental Physiology 2018;103(2):157–69.

[40] Kovalchuk Y, et al. NMDA receptor-mediated subthreshold Ca^{2+} signals in spines of hippocampal neurons. Journal of Neuroscience 2000;20(5):1791–9.

[41] Sabatini BL, Oertner TG, Svoboda K. The life cycle of Ca^{2+} ions in dendritic spines. Neuron 2002;33(3):439–52.

[42] Yuste R, et al. Mechanisms of calcium influx into hippocampal spines: heterogeneity among spines, coincidence detection by NMDA receptors, and optical quantal analysis. Journal of Neuroscience 1999;19(6):1976–87.

[43] Koester HJ, Sakmann B. Calcium dynamics associated with action potentials in single nerve terminals of pyramidal cells in layer 2/3 of the young rat neocortex. The Journal of Physiology 2000;529(3):625–46.

[44] Nevian T, Sakmann B. Spine Ca^{2+} signaling in spike-timing-dependent plasticity. Journal of Neuroscience 2006;26(43):11001–13.

[45] Burnashev N, et al. Fractional calcium currents through recombinant GluR channels of the NMDA, AMPA and kainate receptor subtypes. The Journal of Physiology 1995; 485(2):403–18.

[46] Garaschuk O, et al. Fractional Ca^{2+} currents through somatic and dendritic glutamate receptor channels of rat hippocampal CA1 pyramidal neurones. The Journal of Physiology 1996;491(3):757–72.

[47] Rogers M, Dani JA. Comparison of quantitative calcium flux through NMDA, ATP, and ACh receptor channels. Biophysical Journal 1995;68(2):501–6.

[48] Schneggenburger R, et al. Fractional contribution of calcium to the cation current through glutamate receptor channels. Neuron 1993;11(1):133–43.

[49] Skeberdis VA, et al. Protein kinase A regulates calcium permeability of NMDA receptors. Nature Neuroscience 2006;9(4):501–10.

[50] Sobczyk A, Svoboda K. Activity-dependent plasticity of the NMDA-receptor fractional Ca^{2+} current. Neuron 2007;53(1):17–24.

[51] Mayer ML, Westbrook GL, Guthrie PB. Voltage-dependent block by Mg^{2+} of NMDA responses in spinal cord neurones. Nature 1984;309(5965):261–3.

[52] Nowak L, et al. Magnesium gates glutamate-activated channels in mouse central neurones. Nature 1984;307(5950):462–5.

[53] Sobczyk A, Scheuss V, Svoboda K. NMDA receptor subunit-dependent Ca^{2+} signaling in individual hippocampal dendritic spines. Journal of Neuroscience 2005;25(26): 6037–46.

[54] Clem RL, Barth A. Pathway-specific trafficking of native AMPARs by in vivo experience. Neuron 2006;49(5):663–70.

[55] Tóth K, McBain CJ. Afferent-specific innervation of two distinct AMPA receptor subtypes on single hippocampal interneurons. Nature Neuroscience 1998;1(7):572–8.

[56] Ferraguti F, Shigemoto R. Metabotropic glutamate receptors. Cell and Tissue Research 2006;326(2):483–504.

[57] Lüscher C, Huber KM. Group 1 mGluR-dependent synaptic long-term depression: mechanisms and implications for circuitry and disease. Neuron 2010;65(4):445–59.

[58] Hartmann J, et al. TRPC3 channels are required for synaptic transmission and motor coordination. Neuron 2008;59(3):392–8.

[59] Koch C, Segev I. Methods in neuronal modeling: from ions to networks. MIT Press; 1998.

[60] Ohmiya Y, Hirano T. Shining the light: the mechanism of the bioluminescence reaction of calcium-binding photoproteins. Chemistry & Biology 1996;3(5):337–47.

[61] Cobbold PH, Rink T. Fluorescence and bioluminescence measurement of cytoplasmic free calcium. Biochemical Journal 1987;248(2):313.

[62] Bakayan A, et al. Red fluorescent protein-aequorin fusions as improved bioluminescent Ca^{2+} reporters in single cells and mice. PLoS One 2011;6(5).

[63] Xu X, et al. Imaging protein interactions with bioluminescence resonance energy transfer (BRET) in plant and mammalian cells and tissues. Proceedings of the National Academy of Sciences 2007;104(24):10264–9.

[64] Shimomura O, Kishi Y, Inouye S. The relative rate of aequorin regeneration from apoaequorin and coelenterazine analogues. Biochemical Journal 1993;296(3):549–51.

[65] Grynkiewicz G, Poenie M, Tsien RY. A new generation of Ca^{2+} indicators with greatly improved fluorescence properties. Journal of Biological Chemistry 1985;260(6):3440–50.

[66] Tsien RY. Fluorescent probes of cell signaling. Annual Review of Neuroscience 1989;12(1):227–53.

[67] Tsien R, Rink T, Poenie M. Measurement of cytosolic free Ca^{2+} in individual small cells using fluorescence microscopy with dual excitation wavelengths. Cell Calcium 1985;6(1–2):145–57.

[68] Chang Liao M-L, et al. Sensing cardiac electrical activity with a cardiac myocyte–targeted optogenetic voltage indicator. Circulation Research 2015;117(5):401–12.

[69] Quinn TA, et al. Electrotonic coupling of excitable and nonexcitable cells in the heart revealed by optogenetics. Proceedings of the National Academy of Sciences 2016;113(51):14852–7.

[70] Nagai T, et al. Expanded dynamic range of fluorescent indicators for Ca^{2+} by circularly permuted yellow fluorescent proteins. Proceedings of the National Academy of Sciences 2004;101(29):10554–9.

[71] Jares-Erijman EA, Jovin TM. FRET imaging. Nature Biotechnology 2003;21(11):1387–95.

[72] Chalasani SH, et al. Dissecting a circuit for olfactory behaviour in *Caenorhabditis elegans*. Nature 2007;450(7166):63–70.

[73] Dombeck DA, et al. Functional imaging of hippocampal place cells at cellular resolution during virtual navigation. Nature Neuroscience 2010;13(11):1433.

[74] Fletcher ML, et al. Optical imaging of postsynaptic odor representation in the glomerular layer of the mouse olfactory bulb. Journal of Neurophysiology 2009;102(2):817–30.

[75] Wang JW, et al. Two-photon calcium imaging reveals an odor-evoked map of activity in the fly brain. Cell 2003;112(2):271–82.

[76] Nakai J, Ohkura M, Imoto K. A high signal-to-noise Ca^{2+} probe composed of a single green fluorescent protein. Nature Biotechnology 2001;19(2):137−41.

[77] Tian L, et al. Imaging neural activity in worms, flies and mice with improved GCaMP calcium indicators. Nature Methods 2009;6(12):875.

[78] Mank M, Griesbeck O. Genetically encoded calcium indicators. Chemical Reviews 2008;108(5):1550−64.

[79] Oliver AE, et al. Effects of temperature on calcium-sensitive fluorescent probes. Biophysical Journal 2000;78(4):2116−26.

[80] Helmchen F, Borst J, Sakmann B. Calcium dynamics associated with a single action potential in a CNS presynaptic terminal. Biophysical Journal 1997;72(3):1458−71.

[81] Neher E, Augustine G. Calcium gradients and buffers in bovine chromaffin cells. The Journal of Physiology 1992;450(1):273−301.

[82] Helmchen F, Imoto K, Sakmann B. Ca^{2+} buffering and action potential-evoked Ca^{2+} signaling in dendrites of pyramidal neurons. Biophysical Journal 1996;70(2): 1069−81.

[83] Regehr WG, Tank DW. Dendritic calcium dynamics. Current Opinion in Neurobiology 1994;4(3):373−82.

[84] Sullivan MR, et al. In vivo calcium imaging of circuit activity in cerebellar cortex. Journal of Neurophysiology 2005;94(2):1636−44.

[85] Ohki K, et al. Functional imaging with cellular resolution reveals precise microarchitecture in visual cortex. Nature 2005;433(7026):597−603.

[86] Li Y, et al. Experience with moving visual stimuli drives the early development of cortical direction selectivity. Nature 2008;456(7224):952−6.

[87] Greenberg DS, Houweling AR, Kerr JN. Population imaging of ongoing neuronal activity in the visual cortex of awake rats. Nature Neuroscience 2008;11(7):749.

[88] Rochefort NL, et al. Development of direction selectivity in mouse cortical neurons. Neuron 2011;71(3):425−32.

[89] Sumbre G, et al. Entrained rhythmic activities of neuronal ensembles as perceptual memory of time interval. Nature 2008;456(7218):102−6.

[90] Wachowiak M, Denk W, Friedrich RW. Functional organization of sensory input to the olfactory bulb glomerulus analyzed by two-photon calcium imaging. Proceedings of the National Academy of Sciences 2004;101(24):9097−102.

[91] Dombeck DA, Graziano MS, Tank DW. Functional clustering of neurons in motor cortex determined by cellular resolution imaging in awake behaving mice. Journal of Neuroscience 2009;29(44):13751−60.

[92] Oka Y, et al. Odorant receptor map in the mouse olfactory bulb: in vivo sensitivity and specificity of receptor-defined glomeruli. Neuron 2006;52(5):857−69.

[93] Galizia CG, et al. The glomerular code for odor representation is species specific in the honeybee *Apis mellifera*. Nature Neuroscience 1999;2(5):473−8.

[94] Wachowiak M, Cohen LB. Representation of odorants by receptor neuron input to the mouse olfactory bulb. Neuron 2001;32(4):723−35.

[95] Brustein E, et al. "In vivo" monitoring of neuronal network activity in zebrafish by two-photon Ca^{2+} imaging. Pflügers Archiv 2003;446(6):766−73.

[96] Svoboda K, et al. In vivo dendritic calcium dynamics in neocortical pyramidal neurons. Nature 1997;385(6612):161−5.

[97] Sohya K, et al. GABAergic neurons are less selective to stimulus orientation than excitatory neurons in layer II/III of visual cortex, as revealed by in vivo functional Ca^{2+} imaging in transgenic mice. Journal of Neuroscience 2007;27(8):2145−9.

[98]   Stosiek C, et al. In vivo two-photon calcium imaging of neuronal networks. Proceedings of the National Academy of Sciences 2003;100(12):7319–24.

[99]   Sato TR, et al. The functional microarchitecture of the mouse barrel cortex. PLoS Biology 2007;5(7).

[100] Demarque M, Spitzer NC. Activity-dependent expression of Lmx1b regulates specification of serotonergic neurons modulating swimming behavior. Neuron 2010;67(2): 321–34.

[101] Takano T, et al. Astrocyte-mediated control of cerebral blood flow. Nature Neuroscience 2006;9(2):260–7.

[102] Yaksi E, et al. Transformation of odor representations in target areas of the olfactory bulb. Nature Neuroscience 2009;12(4):474.

[103] Nagayama S, et al. In vivo simultaneous tracing and Ca^{2+} imaging of local neuronal circuits. Neuron 2007;53(6):789–803.

[104] Yu D, et al. Detection of calcium transients in Drosophila mushroom body neurons with camgaroo reporters. Journal of Neuroscience 2003;23(1):64–72.

[105] Hasan MT, et al. Functional fluorescent Ca^{2+} indicator proteins in transgenic mice under TET control. PLoS Biology 2004;2(6).

[106] Li J, et al. Early development of functional spatial maps in the zebrafish olfactory bulb. Journal of Neuroscience 2005;25(24):5784–95.

[107] Díez-García J, et al. Activation of cerebellar parallel fibers monitored in transgenic mice expressing a fluorescent Ca^{2+} indicator protein. European Journal of Neuroscience 2005;22(3):627–35.

[108] Seelig JD, et al. Two-photon calcium imaging from head-fixed Drosophila during optomotor walking behavior. Nature Methods 2010;7(7):535.

[109] Lütcke H, et al. Optical recording of neuronal activity with a genetically-encoded calcium indicator in anesthetized and freely moving mice. Frontiers in Neural Circuits 2010;4:9.

[110] Horikawa K, et al. Spontaneous network activity visualized by ultrasensitive Ca^{2+} indicators, yellow Cameleon-Nano. Nature Methods 2010;7(9):729.

[111] Wallace DJ, et al. Single-spike detection in vitro and in vivo with a genetic Ca^{2+} sensor. Nature Methods 2008;5(9):797.

[112] Mank M, et al. A FRET-based calcium biosensor with fast signal kinetics and high fluorescence change. Biophysical Journal 2006;90(5):1790–6.

[113] Heider B, et al. Two-photon imaging of calcium in virally transfected striate cortical neurons of behaving monkey. PLoS One 2010;5(11).

[114] Heim N, et al. Improved calcium imaging in transgenic mice expressing a troponin C–based biosensor. Nature Methods 2007;4(2):127–9.

[115] Mank M, et al. A genetically encoded calcium indicator for chronic in vivo two-photon imaging. Nature Methods 2008;5(9):805.

[116] Nagai T, et al. Circularly permuted green fluorescent proteins engineered to sense Ca^{2+}. Proceedings of the National Academy of Sciences 2001;98(6):3197–202.

[117] Palmer AE, et al. Ca^{2+} indicators based on computationally redesigned calmodulin-peptide pairs. Chemistry & Biology 2006;13(5):521–30.

[118] Barreto-Chang OL, Dolmetsch RE. Calcium imaging of cortical neurons using Fura-2 AM. Journal of Visualized Experiments 2009;(23):e1067.

[119] Helmchen F, Konnerth A. Imaging in neuroscience: a laboratory manual. Cold Spring Harbor Laboratory Press; 2011.

[120] Garaschuk O, et al. Large-scale oscillatory calcium waves in the immature cortex. Nature Neuroscience 2000;3(5):452−9.

[121] Yuste R, Katz LC. Control of postsynaptic Ca^{2+} influx in developing neocortex by excitatory and inhibitory neurotransmitters. Neuron 1991;6(3):333−44.

[122] Yuste R, Peinado A, Katz LC. Neuronal domains in developing neocortex. Science 1992;257(5070):665−9.

[123] Garaschuk O, Hanse E, Konnerth A. Developmental profile and synaptic origin of early network oscillations in the CA1 region of rat neonatal hippocampus. The Journal of Physiology 1998;507(1):219−36.

[124] Feller MB, et al. Requirement for cholinergic synaptic transmission in the propagation of spontaneous retinal waves. Science 1996;272(5265):1182−7.

[125] Jaffe DB, et al. The spread of Na^+ spikes determines the pattern of dendritic Ca^{2+} entry into hippocampal neurons. Nature 1992;357(6375):244−6.

[126] Eilers J, Konnerth A. Dye loading with patch pipettes. Cold Spring Harbor Protocols 2009;2009(4). https://doi.org/10.1101/pdb.prot5201.

[127] Margrie TW, et al. Targeted whole-cell recordings in the mammalian brain in vivo. Neuron 2003;39(6):911−8.

[128] Judkewitz B, et al. Targeted single-cell electroporation of mammalian neurons in vivo. Nature Protocols 2009;4(6):862.

[129] Kitamura K, et al. Targeted patch-clamp recordings and single-cell electroporation of unlabeled neurons in vivo. Nature Methods 2008;5(1):61−7.

[130] Nevian T, Helmchen F. Calcium indicator loading of neurons using single-cell electroporation. Pflügers Archiv-European Journal of Physiology 2007;454(4):675−88.

[131] Garaschuk O, Milos R-I, Konnerth A. Targeted bulk-loading of fluorescent indicators for two-photon brain imaging in vivo. Nature Protocols 2006;1(1):380−6.

[132] Connor JA, et al. Reduced voltage-dependent Ca^{2+} signaling in CA1 neurons after brief ischemia in gerbils. Journal of Neurophysiology 1999;81(1):299−306.

[133] Gelperin A, Flores J. Vital staining from dye-coated microprobes identifies new olfactory interneurons for optical and electrical recording. Journal of Neuroscience Methods 1997;72(1):97−108.

[134] Kreitzer AC, et al. Monitoring presynaptic calcium dynamics in projection fibers by in vivo loading of a novel calcium indicator. Neuron 2000;27(1):25−32.

[135] O'Donovan MJ, et al. Calcium imaging of network function in the developing spinal cord. Cell Calcium 2005;37(5):443−50.

[136] Nagayama S, et al. Differential axonal projection of mitral and tufted cells in the mouse main olfactory system. Frontiers in Neural Circuits 2010;4:120.

[137] Cetin A, et al. Stereotaxic gene delivery in the rodent brain. Nature Protocols 2006;1(6):3166.

[138] Dittgen T, et al. Lentivirus-based genetic manipulations of cortical neurons and their optical and electrophysiological monitoring in vivo. Proceedings of the National Academy of Sciences 2004;101(52):18206−11.

[139] Monahan PE, Samulski RJ. Adeno-associated virus vectors for gene therapy: more pros than cons? Molecular Medicine Today 2000;6(11):433−40.

[140] Lilley CE, et al. Multiple immediate-early gene-deficient herpes simplex virus vectors allowing efficient gene delivery to neurons in culture and widespread gene delivery to the central nervous system in vivo. Journal of Virology 2001;75(9):4343−56.

[141] Osakada F, et al. New rabies virus variants for monitoring and manipulating activity and gene expression in defined neural circuits. Neuron 2011;71(4):617–31.

[142] De Vry J, et al. In vivo electroporation of the central nervous system: a non-viral approach for targeted gene delivery. Progress in Neurobiology 2010;92(3):227–44.

[143] Heim N, Griesbeck O. Genetically encoded indicators of cellular calcium dynamics based on troponin C and green fluorescent protein. Journal of Biological Chemistry 2004;279(14):14280–6.

[144] Pologruto TA, Yasuda R, Svoboda K. Monitoring neural activity and Ca^{2+} with genetically encoded Ca^{2+} indicators. Journal of Neuroscience 2004;24(43):9572–9.

[145] Tsai PS, et al. All-optical histology using ultrashort laser pulses. Neuron 2003;39(1): 27–41.

[146] Dana H, et al. Thy1 transgenic mice expressing the red fluorescent calcium indicator jRGECO1a for neuronal population imaging in vivo. PLoS One 2018;13(10).

[147] Hendel T, et al. Fluorescence changes of genetic calcium indicators and OGB-1 correlated with neural activity and calcium in vivo and in vitro. Journal of Neuroscience 2008;28(29):7399–411.

[148] Andermann ML, Kerlin AM, Reid C. Chronic cellular imaging of mouse visual cortex during operant behavior and passive viewing. Frontiers in Cellular Neuroscience 2010; 4:3.

[149] Bozza T, et al. In vivo imaging of neuronal activity by targeted expression of a genetically encoded probe in the mouse. Neuron 2004;42(1):9–21.

[150] Mao T, et al. Characterization and subcellular targeting of GCaMP-type genetically-encoded calcium indicators. PLoS One 2008;3(3).

[151] Shigetomi E, et al. A genetically targeted optical sensor to monitor calcium signals in astrocyte processes. Nature Neuroscience 2010;13(6):759.

[152] Luo L, Callaway EM, Svoboda K. Genetic dissection of neural circuits. Neuron 2008; 57(5):634–60.

[153] Makeig S, et al. Blind separation of auditory event-related brain responses into independent components. Proceedings of the National Academy of Sciences 1997;94(20): 10979–84.

[154] Mukamel EA, Nimmerjahn A, Schnitzer MJ. Automated analysis of cellular signals from large-scale calcium imaging data. Neuron 2009;63(6):747–60.

[155] Mitra PP, Pesaran B. Analysis of dynamic brain imaging data. Biophysical Journal 1999;76(2):691–708.

[156] Nagel G, et al. Channelrhodopsin-2, a directly light-gated cation-selective membrane channel. Proceedings of the National Academy of Sciences 2003;100(24):13940–5.

[157] Boyden ES, et al. Millisecond-timescale, genetically targeted optical control of neural activity. Nature Neuroscience 2005;8(9):1263–8.

[158] Hou JH, et al. Simultaneous mapping of membrane voltage and calcium in zebrafish heart in vivo reveals chamber-specific developmental transitions in ionic currents. Frontiers in Physiology 2014;5:344.

[159] Dempsey GT, et al. Cardiotoxicity screening with simultaneous optogenetic pacing, voltage imaging and calcium imaging. Journal of Pharmacological and Toxicological Methods 2016;81:240–50.

[160] Kaestner L, et al. Genetically encoded voltage indicators in circulation research. International Journal of Molecular Sciences 2015;16(9):21626–42.

[161] Loew LM, Bonneville GW, Surow J. Charge shift optical probes of membrane potential. Theory. Biochemistry 1978;17(19):4065–71.

[162] Miller EW. Small molecule fluorescent voltage indicators for studying membrane potential. Current Opinion in Chemical Biology 2016;33:74—80.

[163] McIsaac RS, Bedbrook CN, Arnold FH. Recent advances in engineering microbial rhodopsins for optogenetics. Current Opinion in Structural Biology 2015;33:8—15.

[164] Xu Y, Zou P, Cohen AE. Voltage imaging with genetically encoded indicators. Current Opinion in Chemical Biology 2017;39:1—10.

[165] Jia Z, et al. Stimulating cardiac muscle by light: cardiac optogenetics by cell delivery. Circulation: Arrhythmia and Electrophysiology 2011;4(5):753—60.

[166] Lam P-Y, et al. A high-conductance chemo-optogenetic system based on the vertebrate channel Trpa1b. Scientific Reports 2017;7(1):1—12.

[167] Klimas A, et al. OptoDyCE as an automated system for high-throughput all-optical dynamic cardiac electrophysiology. Nature Communications 2016;7(1):1—12.

[168] Nussinovitch U, Gepstein L. Optogenetics for in vivo cardiac pacing and resynchronization therapies. Nature Biotechnology 2015;33(7):750—4.

[169] Yu J, et al. Cardiac optogenetics: enhancement by all-trans-retinal. Scientific Reports 2015;5:16542.

[170] Feola I, et al. Localized optogenetic targeting of rotors in atrial cardiomyocyte monolayers. Circulation: Arrhythmia and Electrophysiology 2017;10(11):e005591.

[171] Majumder R, et al. Optogenetics enables real-time spatiotemporal control over spiral wave dynamics in an excitable cardiac system. eLife 2018;7:e41076.

[172] Crocini C, et al. Optogenetics design of mechanistically-based stimulation patterns for cardiac defibrillation. Scientific Reports 2016;6:35628.

[173] Scardigli M, et al. Real-time optical manipulation of cardiac conduction in intact hearts. The Journal of Physiology 2018;596(17):3841—58.

[174] Streit J, Kleinlogel S. Dynamic all-optical drug screening on cardiac voltage-gated ion channels. Scientific Reports 2018;8(1):1—13.

[175] Quiñonez Uribe RA, et al. Energy-reduced arrhythmia termination using global photostimulation in optogenetic murine hearts. Frontiers in Physiology 2018;9:1651.

[176] Park SA, et al. Optical mapping of optogenetically shaped cardiac action potentials. Scientific Reports 2014;4:6125.

[177] Li Q, et al. Electrophysiological properties and viability of neonatal rat ventricular myocyte cultures with inducible ChR2 expression. Scientific Reports 2017;7(1):1—12.

[178] Wang Y, et al. Optogenetic control of heart rhythm by selective stimulation of cardiomyocytes derived from Pnmt$^+$ cells in murine heart. Scientific Reports 2017;7(1):1—10.

[179] Zaglia T, et al. Optogenetic determination of the myocardial requirements for extrasystoles by cell type-specific targeting of ChannelRhodopsin-2. Proceedings of the National Academy of Sciences 2015;112(32):E4495—504.

[180] Watanabe M, et al. Optogenetic manipulation of anatomical re-entry by light-guided generation of a reversible local conduction block. Cardiovascular Research 2017;113(3):354—66.

[181] Björk S, et al. Evaluation of optogenetic electrophysiology tools in human stem cell-derived cardiomyocytes. Frontiers in Physiology 2017;8:884.

[182] Beacher J. LEDs for fluorescence microscopy. Photonics Spectra 2008;42(3):13.

[183] Schmieder F, et al. Holographically generated structured illumination for cell stimulation in optogenetics. In: Digital optical technologies 2017. International Society for Optics and Photonics; 2017.

[184] Arrenberg AB, et al. Optogenetic control of cardiac function. Science 2010;330(6006): 971−4.

[185] Prando V, et al. Dynamics of neuroeffector coupling at cardiac sympathetic synapses. The Journal of Physiology 2018;596(11):2055−75.

[186] Klimas A, Entcheva EG. Toward microendoscopy-inspired cardiac optogenetics in vivo: technical overview and perspective. Journal of Biomedical Optics 2014; 19(8):080701.

[187] Jaimes III R, et al. A technical review of optical mapping of intracellular calcium within myocardial tissue. American Journal of Physiology - Heart and Circulatory Physiology 2016;310(11):H1388−401.

[188] Adrian RH, et al. Reviews of physiology, biochemistry and pharmacology, vol. 79. Springer Science & Business Media; 2006.

[189] Cohen L, et al. Changes in axon fluorescence during activity: molecular probes of membrane potential. Journal of Membrane Biology 1974;19(1):1−36.

[190] Hochbaum DR, et al. All-optical electrophysiology in mammalian neurons using engineered microbial rhodopsins. Nature Methods 2014;11(8):825.

[191] Petreanu L, et al. The subcellular organization of neocortical excitatory connections. Nature 2009;457(7233):1142−5.

[192] Zhang H, Reichert E, Cohen AE. Optical electrophysiology for probing function and pharmacology of voltage-gated ion channels. eLife 2016;5:e15202.

[193] Benninger RK, Piston DW. Two-photon excitation microscopy for the study of living cells and tissues. Current Protocols in Cell Biology 2013;59(1):4.11. 1−4.11. 24.

[194] Yang W, et al. Simultaneous two-photon imaging and two-photon optogenetics of cortical circuits in three dimensions. eLife 2018;7:e32671.

[195] Svoboda K, Yasuda R. Principles of two-photon excitation microscopy and its applications to neuroscience. Neuron 2006;50(6):823−39.

[196] Calvo M, Villalobos C, Núñez L. Calcium imaging in neuron cell death. In: Neuronal cell death. Springer; 2015. p. 73−85.

Electronic circuits in patch-clamp system

6.1 Introduction

Technological advancements facilitated the patch-clamp system becoming more user-friendly and commercially available. Although it has become easier to work with the new patch-clamp systems, the need for evaluating the validity of results and analysis is more critical nowadays. For instance, in traditional analog patch-clamp systems, all compensations of series resistances, whole-cell capacitance, and pipette offset had to be adjusted manually, while in the modern digital patch-clamp systems, adjustment is automatic. This chapter provides users of the automated patch-clamp system with deep insight into the application of all the circuits that have been implemented in the system, so they can have a better understanding of the reliability and validity of their experimental results. Although basic designs of voltage-clamp circuits can be found in details by Strickhalm [1], Smith et al. [2], and Sigworth [3], in this chapter bioamplifier circuit, compensation circuits, filters, and data acquisition unit have been discussed in detail.

The aim of this chapter is to present basic construction, calibration, and validation of a patch-clamp system that will help scientists in different fields of medicine and engineering to design, validate, and conduct experiments. With recent updates to computer software, digital signal processing, and electronic circuits, the engineering perspective of updated versions has become mandatory to understand basic electronics of the patch-clamp so that researchers can fully use the patch-clamp technique. It is expected that learning different circuits and elements of the electrophysiology setup and providing basic knowledge of patch-clamp electronics will make scientists more comfortable and confident working with modern patch-clamp systems. This will enable them to design their own patch-clamp systems, obtain core experience to assemble and modify commercial patch-clamp systems that meet specific requirements for their projects, and have a training step for using this system.

Fig. 6.1 shows a general block diagram for a patch-clamp system. It includes a bioamplifier and buffer blocks for amplifying signals from the cell and the digital-to-analog convertor or signal generator to stimulate the cell under the patch. In addition, a filter stage is needed to remove noises from the bioamplifier's output and be prepared for the analog-to-digital convertor. The analog-to-digital

Electrophysiology Measurements for Studying Neural Interfaces. https://doi.org/10.1016/B978-0-12-817070-0.00006-3
143

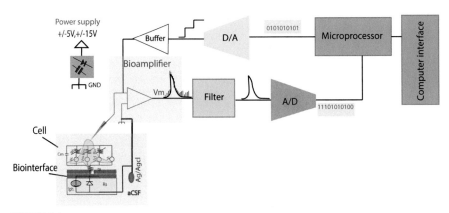

FIGURE 6.1

Schematic of a block diagram for patch-clamp system consists of a bioamplifier, filter, A//D, D/A, buffer, microprocessor, computer interface, and power supply. It also shows that a grown cell on top of a biointerface is approached, and there is an electrical equivalent circuit of the cell and the interface, which are integrated.

convertor (A/D) is used to convert analog signals to digital form so that they can be recorded and processed by a microprocessor unit. The microprocessor is used also to program the digital-to-analog convertor or signal generator to produce a desired voltage and current signal for voltage and current-clamp experiments, and finally, the power supply that supplies all blocks with electrical power. All the aforementioned circuits will be discussed in terms of their role in the patch-clamp system.

6.2 Power supply

The amplifier and other elements of the patch-patch-clamp system are powered by a regulated ±15 and ±5 V power supply (see Fig. 6.2) with a minimum ripple of voltage (<0.2 mv). Generally, this power supply consists of the input transformer to convert 240 to 18 V, a rectifier circuit made of diodes (to convert AC signal to DC), capacitance filters (to reduce ripples from the output of rectifier), voltage regulators (to keep the output at the constant ±15 V), and filter capacitors to reduce noise.

Grounding and Faraday cage: One of the most important factors in recording high-quality signals and accurate measurements is considering grounding and noise cancellation approaches in electronic circuits. The most disturbing noise is 50/60 Hz, from the power cord; it is high enough to kill the patch-clamp measurements. In this regard, different filtering elements are used in the circuits to avoid ground loops in circuits and connect the body of the equipment to the ground.

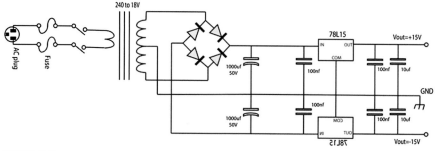

FIGURE 6.2

Schematic of a power supply circuit. This circuit include rectifier diodes and filter capacitances at its output. Next, voltage regulators of 78L15 are for regulating the output at the fixed DC level of 15 V.

Along with the electronic elements, using a Faraday cage is very promising for reducing the noise level to a minimum, as there are always electromagnetic noises with different frequency spectra and they can be easily induced at the head stage of the patch-clamp and Ag/AgCl electrode.

Input circuit: Signal generator or digital-to-analog interface provides command voltages with different holding potentials for the voltage or current-clamp circuits through the input stage (see Fig. 6.3). The input stage includes two different inputs which go through operational amplifier (op-amp) (U10 and U11), and an optional (VH) holding potential can be applied to U12 at the end. VH and its polarity can be selected by switch S_3. The S_1 and S_2 switches can change the amplifier function to a negative or positive amplifier for the input command voltage. The input circuit is also connected to ground voltage (V_g) and junction zero (Pz) potentiometer polarity selector S_4 for input offset corrections [4].

6.3 Signal generators

Digital-to-analog converters (D/A): D/A IC modules are an inevitable part of signal generators. The input of DACs are digital codes; they are converted to analog format at the output. Fig. 6.4A and B shows the schematic of n-bit DAC with digital input and the simple time domain of analog signal at the output. As shown in Fig. 6.4, the output looks like discontinuous pulses that after filtering convert to an analog format. The filter can be used to remove high-frequency harmonics, resulting in a smooth desired frequency. The resolution of DACs depends on the number of bits used in the conversion process; therefore, the higher the number of bits we have, the higher the resolution we can obtain.

Direct digital synthesizer (DDS): An internal programmable signal generator is used in the patch-clamp system (see Fig. 6.5). DDS ICs consist of a DAC

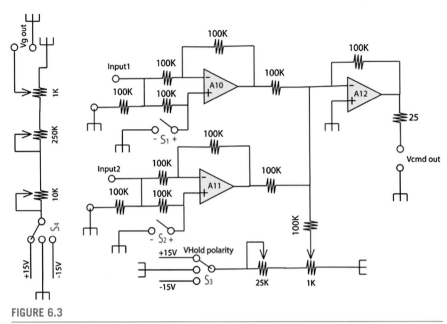

FIGURE 6.3

Schematic of input circuit. Input circuit consists of two inputs that can be used for inverted mode by S_1 and S_2, and the polarity of an offset holding potential can be chosen by S_3. Finally, the output which will be the command voltage will be obtained after amplification through U12.

Reproduced from Rouzrokh A, Ebrahimi SA, Mahmoudian M. Construction, calibration, and validation of a simple patch-clamp amplifier for physiology education. Advances in Physiology Education 2009;33(2):121–9.

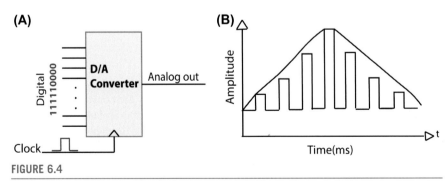

FIGURE 6.4

(A) Block diagram of a digital-to-analog converter. (B) Converted signal at the analog output of D/A block diagram.

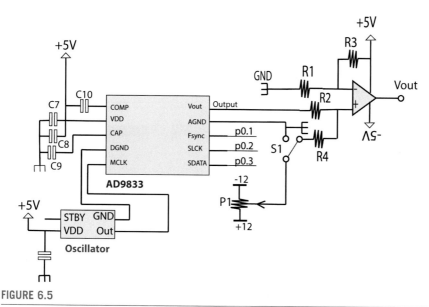

FIGURE 6.5

Schematic of a DDS-based signal generator. The clock source of this programmable DDS is supported by a crystal oscillator.

unit, and this pulse generator consists of a DDS, which is programmed via a microprocessor or microcontroller. This wave generator produces different wave shapes such as sine, square, and triangular with frequencies up to 2 MHz with 0.1 Hz resolution. The output amplitude of the pulse generator can be boosted by an extra opamp circuit.

The registers of the DDS chip for the desired output signal are programmed through a microprocessor/controller by three pins of Fsync, SLCK, and SDATA based on the protocol explained in the product (AD9830) datasheet [5].

6.4 Head stage

The head stage circuit is a current-to-voltage convertor (see Fig. 6.6) utilizing a high-input impedance op-amp. This bioamplifier has a unit gain and measures the tip of the electrode through its noninverting input. The feedback resistor (R_f) is usually in the range of 100 MOhm for whole-cell experiments. Although it depends on the level of current injection to the cell and the current recording level, this resistance can be changed. A 25 Ohm resistor decouples the output from the cable capacitance to prevent oscillations. In addition, a variable resistance P1 is used for offset adjustment.

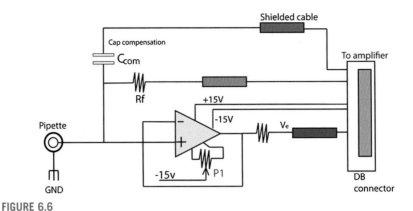

FIGURE 6.6

Schematic of the head stage circuit. This circuit consists of a unity gain amplifier that plays the role of a buffer between the pipette and the input of the voltage-clamp circuit. C_{com} is the capacitance belonging to the negative capacitance compensation circuit. Ionic current passing through feedback resistance of R_f can be measured in the voltage-clamp circuit.

Reproduced from Rouzrokh A, Ebrahimi SA, Mahmoudian M. Construction, calibration, and validation of a simple patch-clamp amplifier for physiology education. Advances in Physiology Education 2009;33(2):121−9.

6.5 Bioamplifier

The patch-clamp amplifier is the core element that can operate in both current-clamp and voltage-clamp mode. In voltage-clamp mode, cell membrane potential is set at an interested voltage while the current passing through the membrane is recorded. The voltage-clamp circuit generally is employed as a feedback system which compares the command voltage with the recorded potential from the membrane. This feedback circuit tries to follow membrane potential and lock it in the applied command voltage, while the deviation of the recorded drop across the membrane from the command voltage is compensated by injecting an external current I_j through the circuit.

This current indicates the ionic current under investigation. It is correlated to the number of the open ion channels in the membrane for the V_c. This small ionic current passes through R_f (feedback resistor) and drops a significant measurable voltage at the output of the amplifier. The dropped voltage due to tip resistance is very small because of low resistance of the electrode (usually 10 MOhm). The change in V_e has been shown to be as follows [1]:

$$\Delta V_e = \theta \times \Delta V_m \tag{6.1}$$

$$\theta = R_m/(R_e + R_m) \tag{6.2}$$

where R_m is the membrane resistance and R_e is the electrode's series resistance. For R_c considerably smaller than R_m, like in patch-clamp electrodes, θ is very small and V_e is very close to V_m. For this reason, there is usually no need for series resistance compensation [1,6].

6.6 Voltage-clamp mode circuit

Membrane potential (V_m) is measured by U0 in head stage and is sent to U1 op-amp through mode switch and S_1 (see Fig. 6.7). By selecting S_1, S_2, and S_3, the amplifier will be in voltage-clamp mode (V_c), electrometer (V_m) mode, or current-clamp mode (CC). In voltage-clamp mode upon applying a command voltage (V_c), a required current (I_j) is injected through U1, which results in a shift of V_e toward V_c. Output of U1 changes in response to V_c as follows:

$$\Delta V_o = \Delta V_e + [(\Delta I_f)(R_f)] \tag{6.3}$$

The current passing through R_f, which is indicating the interested ionic current and conductance, is measured by subtracting output voltage of op-amp U2B from the V_e.

$$\Delta V_i = \Delta(V_o - V_e) = [(\Delta I_f)(R_f)] \tag{6.4}$$

where V_i is applied to the nulling circuit (see Fig. 6.7) for further analysis. Amplified signals of V_e and V_c are provided by U4 and U5 is applied to the electrode input for the input capacitance compensation.

6.7 Current-clamp mode circuit

In current-clamp mode, switch configuration is selected on current-clamp mode by mode switch and S_3 (see Fig. 6.7). In this mode, the membrane potential (direct actual measurement without reading the drop of voltage across another resistance) is recorded through V_e. By applying a command V_c, a constant current passes through the R_f and other elements attached to the tip (microelectrode, cell, ground, etc.). The current can be obtained as follows:

$$I_e = \frac{-V_e}{R_f} \tag{6.5}$$

6.8 Capacitance compensation circuit

There are two types of currents that pass through the membrane: ionic current, which flows through ion channels of interest, and an early capacitance current such as activation of Na^+ channels. In ideal conditions, capacitive currents are linear and not voltage-dependent, and they can be canceled with injection of an external current to the pipette by a capacitor (C_{com}) in the head stage (see Fig. 6.7). With two controls in amplifier circuit, we can compensate parasitic capacitances: (1) negative capacitance circuit, and (2) step function response driven by U4 and U5 op-amps (see Fig. 6.7). U4 is for compensation of the pipette capacitance, and U5 is driven by V_e for cancellation of slower transients (P1) with negative capacitance. The ratio of compensation can be adjusted by using variable resistance of P3.

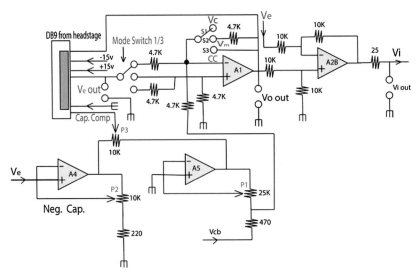

FIGURE 6.7

Schematic of voltage-clamp circuit. Modes of voltage clamp (V_c), current clamp (C_c), and voltmeter (V_m) are selected by S_1, S_2, and S_3 switches. DB9 is the input connection from the head stage. U4, P3, P2, and 220o hm resistance are part of the negative capacitance compensation circuit.

Reproduced from Rouzrokh A, Ebrahimi SA, Mahmoudian M. Construction, calibration, and validation of a simple patch-clamp amplifier for physiology education. Advances in Physiology Education 2009;33(2):121−9.

6.9 Output nulling circuit

In Fig. 6.8, SW2 connects V_e to U8 in current-clamp mode and V_i to U8 in voltage-clamp mode through 25 kOhm calibration resistors; these are explained in the section on calibration. The calibration signal of V_{cb} (proportional to V_c) is applied through U7 and U8 for determining the membrane and electrode resistance (measured by variable resistance of P1) by nulling output signal of the amplifier in current-clamp mode. In voltage-clamp mode, P1 (a variable resistance with dial) is used for leak current cancellation, and P2 for nulling V_m (membrane potential). U8 is an op-amp whose output gain is adjusted by the switch SW3 to 1X or 10X for amplification of the output. Finally, the signal will be filtered by the last stage with a proper filter for a desired cutoff frequency.

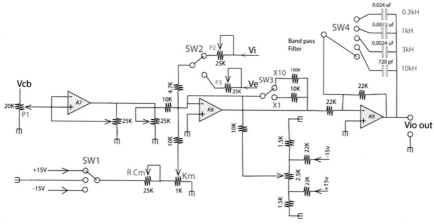

FIGURE 6.8

Schematic of the nulling circuit. This circuit includes three sections, U7 buffer for the calibration of V_{cb}, U8 with all its resistors is the amplifier circuit to boost the V_i signal in case of voltage-clamp or V_e signal in case of current-clamp mode, and U9 plus capacitors and its resistors are the components of output filter. V_{io} out will go to the D/A card.

Reproduced from Rouzrokh A, Ebrahimi SA, Mahmoudian M. Construction, calibration, and validation of a simple patch-clamp amplifier for physiology education. Advances in Physiology Education 2009;33(2):121−9.

6.10 Analog to digital convertor unit

Analog-to-digital convertors (A/D) convert analog signals to digital formats, so that they can be recorded and processed by a computer. Generally, A/D unit samples an input analog signal with n-bit resolution and a sampling rate. For example, AD7768-1 is a 24-bit A/D from *Analog* Devices Inc. It uses 3.3 V reference voltage and sample with $3.3V/2^{24} = 0.196$ uV resolution with sampling rate of up to 2×10^5 samples/second. Fig. 6.9 shows a 24-bit A/D unit and its digital output sampled from the input analog signal. This resolution is high enough to sample amplified signals in the range of 100 uV and mV range. In addition, thanks to fast frequency response of this A/D, it enables to capture fast events and spikes in submillisecond.

6.11 Model cell

A model cell consisting of electrical components is used for initial and periodic calibration of the patch-clamp system (Fig. 6.10). The model cell provided represents a resistance electrode connected to a cell with an access resistance and an access capacitance. Fig. 6.10A shows the equivalent electrical circuits for giga-ohm seal (a), bath or aCSF solution (c), and a whole cell (e) configuration. For a giga-ohm

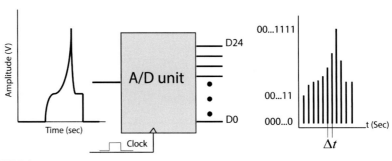

FIGURE 6.9

Schematic of 24-bit A/D unit and its function. An analog input is converted to n-bit digital output with a sampling rate of 1/dt and n-bit resolution.

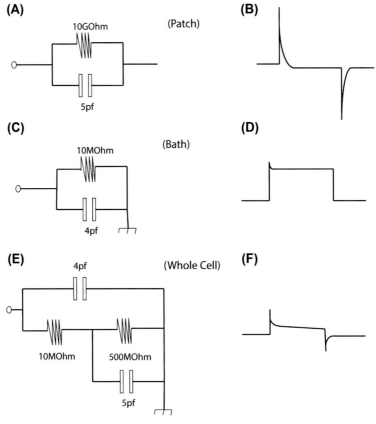

FIGURE 6.10

Electrical equivalent of different elements in the cell mode and their pulse test response. (A) A simple RC model for giga-ohm seal. (B) Pulse test response to the patch model for giga-ohm seal. (C) RC electrical circuit for the batch aCSF medium. (D) Pulse test response to the bath model. (E) Equivalent electrical circuit for the cell. (F) Pulse test response to the cell model.

seal, we observe high transient capacitive currents (see Fig. 6.10B), while after formation of the whole seal, the electrical resistance is reduced from 10 GOhm to several 100 MOhm, and the pulse test results in more resistive behavior (see Fig. 6.10F).

6.12 Calibration and operation tests

To calibrate the patch-clamp system, it is important to check the noise level and the quality of the output signal in voltage-clamp mode. To do that, we can set the input voltage as zero (GND) and then record current (in continuous mode vs. time), while the tip's electrode has been immersed in the aCSF solution. The level of recorded current should be in several picoamp in ideal case when all grounding and proper shielding have been carefully considered, but there are always many noises in the environment that can induce background signals, especially when something is wrong with electrical grounding or shielding. After recording the output current, the frequency spectrum of the time domain signal can be obtained through Fourier transform functions to detect the magnitude of noise in the system. Finally, if the magnitude of noise especially 50/60 Hz was higher than 20 pA; one can perform troubleshooting of the system through short-circuit probe tests.

Composition of intracellular and extracellular solution: To perform whole-cell recording, extracellular solution (aCSF) is prepared by dissolving 10 mM of 4-(2-hydroxyethyl)-1-piperazineethanesulfonic acid, 10 mM of glucose, 2 mM $CaCl_2$, 140 mM of NaCl, 1 mM of $MgCl_2$, 3 mM of KCl, and NaOH (To adjust PH = 7.4) into the distilled water. The patch pipette is filled with (in mM) 140 mM KCl, 2 $MgCl_2$, 10 HEPE, 10 EGTA, 2 Mg-ATP, and pH 7.2 with KOH. Also, pulled patch pipettes (8−12 MOhm) can be used to get a stable giga-ohm seal from the patched cells.

Osmolality: The normal osmolarity for the external solution should be between 300 and 330 mOsm. If the internal solution is made slightly hypoosmotic (by about 10%) in comparison with the external solution, the whole-cell recording could last longer.

Junction potential measurement: The aCSF solution is grounded by an Ag/AgCl electrode. There is an offset in the input of the amplifiers, and to remove it, the ground offset should be adjusted. In normal patch-clamp experiments, the junction potential should be about several millivolts. With similar aCSF medium and proper Ag/AgCl coating, the ground offset should be near zero. We can set the junction potential to zero by adjusting $P_{JUNC\ ZERO}$ in Fig. 6.3.

Linearity test: Next step is using cell model and applying different input voltages to record I-V for the standard equivalent circuits. Therefore, we will be able to check the electronic circuit until the tip after head stage. To ensure linear response of the amplifiers without any rectification behavior, an I-V test (scanning command voltage from −100 to +100 mV and reverse) is given to the cell models.

Recording procedure for whole cell recording: The glass capillary pipette tip is immersed in aCSF bath solution and offset potential compensation as it has been discussed is applied before patching the cell. In voltage-clamp mode, a hyperpolarization step signal is applied to observe the resistance (current) response. Capacitance compensation is adjusted to reduce transient spikes using negative capacitance compensation. After patching the cell and getting giga-ohm seal (as it was discussed in Chapter 2), voltage scan is applied to read I-V characteristic of the cell to find ionic conductances and membrane resting potential. After recording I-V, the S is selected in current-clamp mode to measure membrane potential and action potential generations for different injecting currents.

6.13 Extracellular amplifier circuit

Extracellular recording circuit is similar to a patch-clamp system consisting of two stages of amplifiers and two stages of filters (see Fig. 6.11). It is generally simpler than a patch-clamp circuit, and we can only record extracellular potentials, neural activities, and spikes. In the first stage, a differential amplifier (such as

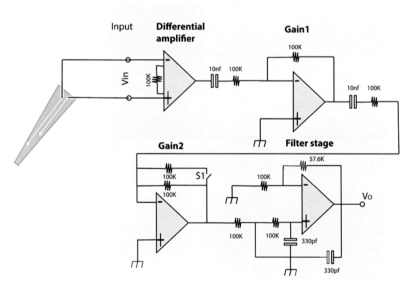

FIGURE 6.11

Schematic of extracellular recording amplifier. It includes a differential amplifier, two stage of voltage amplifiers, and two stages of filters (a low-pass filter and a notch filter for 60 Hz noise).

Land BR, Wyttenbach RA, Johnson BR. Tools for physiology labs: an inexpensive high-performance amplifier and electrode for extracellular recording. Journal of Neuroscience Methods 2001;106(1):47–55.

AD620) is used to amplify the subtraction of input terminals with high signal/noise ratio (high the common-mode rejection ratio). This amplifier is important as it cancels the similar noises on the both terminals effectively. After that two stages of amplification and high-pass filtering, high-pass cutoff of 160 Hz is used to boost input signals and to reduce low frequency noises. Total amplification gain of 100 up to 1000 can be achieved from all different stages.

For the gain below 10 at each stage, the bandwidth of the circuit could be increased, which enables the circuit to respond faster. Filter stage 1 is a low-pass filter designed for a flatness of frequency response and reducing high frequency noises with a cutoff frequency of 5 kHz. Filter stage2 is a notch filter to cancel 60 Hz noise if it is not already eliminated at the first stage.

6.14 Conclusion

This chapter described different blocks of the patch-clamp system, and the designed circuits have been successfully tested by Rouzrokh et al. can be modified or improved based on the new components with higher performance. Building a patch-clamp amplifier is more cost-effective and setting up a patch-clamp rig improves the technical knowledge of researchers and scientists from different backgrounds with regard to electrical engineering perspective of electrophysiology, which is highly demanded and inevitable. Moreover, this chapter also aimed to provide a guideline to scientists for assembling a low-cost system based on their budget.

Appendix 1: Operational amplifier circuits
A1.1 Op-amps

Op-amps are versatile IC chips containing transistor circuits, and we can implement mathematical functions (adders, subtractors, and differentiators) with them. They have two inverting and noninverting inputs and power source terminals. Fig. 6A.1 shows an op-amp with its equivalent electrical circuit containing a voltage-dependent voltage source of gV_{in}. G is the open-loop gain of the op-amp and is different for different op-amps. Generally, in the open-loop circuit, when there are no feedback resistances between the inputs and the output, the gain typically is in the range of 100,000. R_i is the input impedance of the op-amp, which is very high, and for low-bias op-amps, it is in the range of giga Ohm. Therefore, the input current can be considered zero as the current which passes through the feedback loop can be in the range of micro to milliamp, and it is three order to six order higher than the input current. R_o is the output resistance and generally is low, in the range of 50 Ohm.

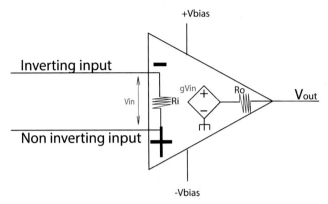

FIGURE 6A.1

Schematic of an operational amplifier (op-amp). Negative and positive terminals are indicating inverting and noninverting terminals respectively. An op-amp can be modeled based on the relationship of the output voltage with the input. Simply a voltage-dependent voltage source which is following input voltage will be at the output with the gV_{in} amplitude and g is the gain of the op-amp.

A1.2 Inverting and noninverting amplifiers

Fig. 6A.2 shows the inverting (A) and noninverting (B) amplifiers, respectively. In the inverting amplifier, the input signal is applied to the inverting terminal and the resulting current can be calculated as follows:

$$I_i = I_f; \; I_i = \frac{V_i}{R_1}; \; I_f = -\frac{V_o}{R_f} \tag{6A1.1}$$

$$\frac{V_o}{V_i} = -\frac{R_2}{R_1} \tag{6A1.2}$$

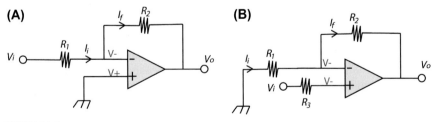

FIGURE 6A.2

Schematic of inverting (A) and noninverting (B) amplifier circuits.

We can reconfigure the amplifier circuit for noninverting amplification by applying input signal to the noninverting terminal and grounding R_1 resistor (Fig. 6A.2B). In this circuit, the output voltage is related to the input as following:

$$V_i = V_+ \tag{6A1.3}$$

$$\frac{V_O - V_+}{R_2} = \frac{V_+}{R_1} \tag{6A1.4}$$

$$\frac{V_{Out}}{V_i} = \frac{R_1 + R_2}{R_1} = \left(1 - \frac{R_2}{R_1}\right) = (1 + G) \tag{6A1.5}$$

where V_+ is the positive terminal of the op-amp, and it is equal to the applied input voltage (V_i) as the terminal current is near zero (it means the potential drop across R_3 is zero). In addition, the total gain of the amplifier from the output voltage to input voltage is equal $1 + G$, which can be defined by selecting R_1 and R_2 values.

A1.3 Adder circuit

Negative adder circuit provides an output signal that is equal to the negative weighted sum of the input signals of V_1 and V_2 (Fig. 6A.3).

$$V_O = -R_3 \times I_f = -\left(\left(\frac{R_3}{R_1}\right)V_1 + \left(\frac{R_3}{R_2}\right)V_2\right) \tag{6A1.6}$$

$$I_f = \frac{V_1}{R_1} + \frac{V_2}{R_2} \tag{6A1.7}$$

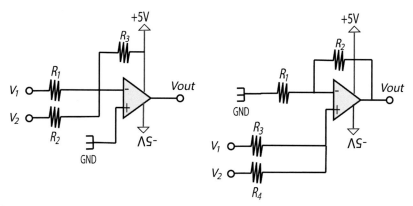

FIGURE 6A.3

Schematic of adder circuits. The left configuration is a negative adder and the right circuit is a positive adder.

A1.4 Negative capacitance circuit

As discussed in Section 6.8, capacitance compensation is very important for the patch-clamp experiments, as it enables us to distinguish the ionic currents that we are interested in to measure from the transient capacitive currents resulting from the cell membrane or tip capacitance. One method for canceling a capacitance is coupling that capacitance with its negative equal capacitance. Negative capacitance circuit facilitates the implementation of a negative capacitance easily. As Fig. 6A.4 illustrates, this circuit is the same as a noninverting amplifier configuration with a capacitance in the feedback path. To understand the equal impedance or capacitance at the input of the circuit, we simply find the relationship of the input current with the input voltage as given in the following equations:

$$I_i = \frac{V_{in} - V_{out}}{Z_{in}} \qquad (6A1.8)$$

Based on the voltage gain (Vout/Vin) in the Section 6A1.5 the input conductance can be obtained as following:

$$\frac{Z_{in}}{V_{in}} = \left(1 - (R_1 - R_2)/R_2\right) = (1 - G) \qquad (6A1.9)$$

Finally, for $G > 1$, we will have a negative value that can be tuned for an effective cancellation of the parasitic impedances such as junction, tip, and membrane capacitances.

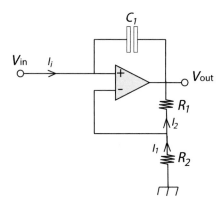

FIGURE 6A.4

Schematic of negative capacitance circuit. This circuit includes a noninverting configuration along with a capacitance in the feedback path.

References

[1] Strickholm A. A single electrode voltage, current-and patch-clamp amplifier with complete stable series resistance compensation. Journal of Neuroscience Methods 1995; 61(1−2):53−66.

[2] Smith TG, Lecar H, Redman SJ. Voltage and patch clamping with microelectrodes. Springer; 2013.

[3] Sigworth F. Electronic design of the patch clamp. In: Single-channel recording. Springer; 1995. p. 95−127.

[4] Rouzrokh A, Ebrahimi SA, Mahmoudian M. Construction, calibration, and validation of a simple patch-clamp amplifier for physiology education. Advances in Physiology Education 2009;33(2):121−9.

[5] VDD, AGND DGND. "Data Sheet AD9833."

[6] Safronov BV, Vogel W. Electrical activity of individual neurons: patch-clamp techniques. In: Modern techniques in neuroscience research. Springer; 1999. p. 173−92.

[7] Land BR, Wyttenbach RA, Johnson BR. Tools for physiology labs: an inexpensive high-performance amplifier and electrode for extracellular recording. Journal of Neuroscience Methods 2001;106(1):47−55.

Index